D1318260

Made in the United States
Text printed on 100%
recycled paper

**Houghton
Mifflin
Harcourt**

Printed in the U.S.A.

ISBN 978-0-544-29550-6

19   0877   19

4500788160          D E F G

Dear Students and Families,

Welcome to **Go Math!**, Grade 5! In this exciting mathematics program, there are hands-on activities to do and real-world problems to solve. Best of all, you will write your ideas and answers right in your book. In **Go Math!**, writing and drawing on the pages helps you think deeply about what you are learning, and you will really understand math!

By the way, all of the pages in your **Go Math!** book are made using recycled paper. We wanted you to know that you can Go Green with **Go Math!**

Sincerely,

The Authors

Made in the United States
Text printed on 100% recycled paper

# GO MATH!

# Authors

**Juli K. Dixon, Ph.D.**
Professor, Mathematics Education
University of Central Florida
Orlando, Florida

**Edward B. Burger, Ph.D.**
President, Southwestern University
Georgetown, Texas

**Steven J. Leinwand**
Principal Research Analyst
American Institutes for
    Research (AIR)
Washington, D.C.

## Contributor

**Rena Petrello**
Professor, Mathematics
Moorpark College
Moorpark, California

**Matthew R. Larson, Ph.D.**
K-12 Curriculum Specialist for
    Mathematics
Lincoln Public Schools
Lincoln, Nebraska

**Martha E. Sandoval-Martinez**
Math Instructor
El Camino College
Torrance, California

## English Language Learners Consultant

**Elizabeth Jiménez**
CEO, GEMAS Consulting
Professional Expert on English
    Learner Education
Bilingual Education and
    Dual Language
Pomona, California

# Fluency with Whole Numbers and Decimals

**Critical Area** Extending division to 2-digit divisors, integrating decimal fractions into the place value system and developing understanding of operations with decimals to hundredths, and developing fluency with whole number and decimal operations

**Real World Project** In the Chef's Kitchen . . . . . . . . . . . . . . . . . . . . **2**

## 1 Place Value, Multiplication, and Expressions 3

**COMMON CORE STATE STANDARDS**

**5.OA Operations and Algebraic Thinking**
**Cluster A** Write and interpret numerical expressions.
5.OA.A.1, 5.OA.A.2

**5.NBT Number and Operations in Base Ten**
**Cluster A** Understand the place value system.
5.NBT.A.1, 5.NBT.A.2

**Cluster B** Perform operations with multi-digit whole numbers and with decimals to hundredths.
5.NBT.B.5, 5.NBT.B.6

✓ Show What You Know . . . . . . . . . . . . . . . . . . . . . . . . 3

Vocabulary Builder . . . . . . . . . . . . . . . . . . . . . . . . . 4

Chapter Vocabulary Cards

Vocabulary Game . . . . . . . . . . . . . . . . . . . . . . . . **4A**

1 Investigate • Place Value and Patterns . . . . . . . . . . **5**
Practice and Homework

2 Place Value of Whole Numbers . . . . . . . . . . . . . . **11**
Practice and Homework

3 Algebra • Properties . . . . . . . . . . . . . . . . . . . . **17**
Practice and Homework

4 Algebra • Powers of 10 and Exponents . . . . . . . . . . **23**
Practice and Homework

5 Algebra • Multiplication Patterns . . . . . . . . . . . . **29**
Practice and Homework

✓ Mid-Chapter Checkpoint . . . . . . . . . . . . . . . . . . **35**

## GO DIGITAL

Go online! Your math lessons are interactive. Use *i*Tools, Animated Math Models, the Multimedia *e*Glossary, and more.

### Chapter 1 Overview

In this chapter, you will explore and discover answers to the following **Essential Questions**:

• How can you use place value, multiplication, and expressions to represent and solve problems?

• How can you read, write, and represent whole numbers through millions?

• How can you use properties and multiplication to solve problems?

• How can you use expressions to represent and solve a problem?

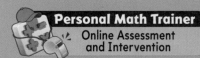

**Personal Math Trainer**
Online Assessment and Intervention

**6** Multiply by 1-Digit Numbers . . . . . . . . . . . . . . . . . . . . **37**
Practice and Homework

**7** Multiply by Multi-Digit Numbers . . . . . . . . . . . . . . . . . **43**
Practice and Homework

**8** Relate Multiplication to Division . . . . . . . . . . . . . . . . . **49**
Practice and Homework

**9** **Problem Solving** • Multiplication and Division . . . . . . . . . . . **55**
Practice and Homework

**10** **Algebra** • Numerical Expressions . . . . . . . . . . . . . . . **61**
Practice and Homework

**11** **Algebra** • Evaluate Numerical Expressions . . . . . . . . . . . . **67**
Practice and Homework

**12** **Algebra** • Grouping Symbols . . . . . . . . . . . . . . . . . **73**
Practice and Homework

✔ Chapter 1 Review/Test . . . . . . . . . . . . . . . . . . . . . . . **80**

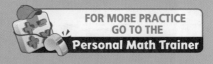

FOR MORE PRACTICE
GO TO THE
**Personal Math Trainer**

**Practice and Homework**

Lesson Check and Spiral Review in every lesson

**Critical Area**

# Fluency with Whole Numbers and Decimals

**Common Core**

**CRITICAL AREA** Extending division to 2-digit divisors, integrating decimal fractions into the place value system and developing understanding of operations with decimals to hundredths, and developing fluency with whole number and decimal operations

Chef preparing lunch in a restaurant

## In the Chef's Kitchen

Restaurant chefs estimate the amount of food they need to buy based on how many diners they expect. They usually use recipes that make enough to serve large numbers of people.

## Get Started  WRITE ▸ Math

Although apples can grow in any of the 50 states, Pennsylvania is one of the top apple-producing states. The ingredients at the right are needed to make 100 servings of Apple Dumplings. Suppose you and a partner want to make this recipe for 25 friends. Adjust the amount of each ingredient to make just 25 servings.

### Important Facts

**Apple Dumplings** (100 servings)
- 100 baking apples
- 72 tablespoons sugar ($4\frac{1}{2}$ cups)
- 14 cups all-purpose flour
- 6 teaspoons baking powder
- 24 eggs
- 80 tablespoons butter (10 sticks of butter)
- 50 tablespoons chopped walnuts ($3\frac{1}{8}$ cups)

**Apple Dumplings** (25 servings)

Completed by _____

## Show What You Know

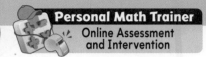

**Personal Math Trainer**
Online Assessment and Intervention

Check your understanding of important skills.

Name _____

▶ **Place Value** **Write the value of each digit for the given number.** (4.NBT.A.1)

1. 2,904

   2 _____

   9 _____

   0 _____

   4 _____

2. 6,423

   6 _____

   4 _____

   2 _____

   3 _____

▶ **Regroup Through Thousands** **Regroup. Write the missing numbers.** (4.NBT.A.1)

3. 40 tens = _____ hundreds

4. 60 hundreds = _____ thousands

5. _____ tens 15 ones = 6 tens 5 ones

6. 18 tens 20 ones = _____ hundreds

▶ **Missing Factors** **Find the missing factor.** (3.OA.A.4)

7. $4 \times$ _____ $= 24$

8. $6 \times$ _____ $= 48$

9. _____ $\times 9 = 63$

## Math in the Real World

Use the clues at the right to find the 7-digit number. What is the number?

### Clues

- This 7-digit number is 8,920,000 when rounded to the nearest ten thousand.
- The digits in the tens and hundreds places are the least and same value.
- The value of the thousands digit is double that of the ten thousands digit.
- The sum of all its digits is 24.

# Vocabulary Builder

▶ **Visualize It** •••••••••••••••••••••••••••••••••••••••••

**Sort the review words into the Venn diagram.**

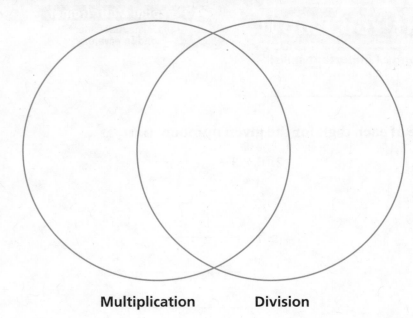

Multiplication        Division

▶ **Understand Vocabulary** ••••••••••••••••••••••••••••••

**Write the preview words that answer the question "What am I?"**

1. I am a group of 3 digits separated by commas in a multidigit

   number. _____

2. I am a mathematical phrase that has numbers and operation signs

   but no equal sign. _____

3. I am operations that undo each other, like multiplication and division.

   _____

4. I am the property that states that multiplying a sum by a
   number is the same as multiplying each addend in the
   sum by the number and then adding the products.

   _____

5. I am a number that tells how many times the base is used

   as a factor. _____

**GO DIGITAL**
• Interactive Student Edition
• Multimedia *eGlossary*

# Chapter 1 Vocabulary

**base**

base

1

**Distributive Property**

propiedad distributiva

17

**evaluate**

evaluar

24

**exponent**

exponente

26

**inverse operations**

operaciones inversas

32

**numerical expression**

expresión numérica

43

**order of operations**

orden de las operaciones

44

**period**

período

49

The property which states that multiplying a sum by a number is the same as multiplying each addend in the sum by the number and then adding the products

Example: $3 \times (4 + 2) = (3 \times 4) + (3 \times 2)$
$(3 \times 6) = 12 + 6$
$18 = 18$

(arithmetic) A number used as a repeated factor
Example: $8^3 = 8 \times 8 \times 8$

base

(geometry) In two dimensions, one side of a triangle or parallelogram that is used to help find the area. In three dimensions, a plane figure, usually a polygon or circle, by which a three-dimensional figure is measured or named

Examples:
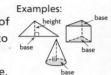

A number that shows how many times the base is used as a factor

exponent

Example: $10^3 = 10 \times 10 \times 10$

To find the value of a numerical or algebraic expression

A mathematical phrase that uses only numbers and operation signs

Example: $(4 + 6) \div 5$

Opposite operations, or operations that undo each other, such as addition and subtraction or multiplication and division

Examples:

| |
|---|
| $6 + 3 = 9$ |
| $9 - 6 = 3$ |

| |
|---|
| $5 \times 2 = 10$ |
| $10 \div 2 = 5$ |

Each group of three digits separated by commas in a multi-digit number

Periods

| MILLIONS | | | THOUSANDS | | | ONES | | |
|---|---|---|---|---|---|---|---|---|
| Hundreds | Tens | Ones | Hundreds | Tens | Ones | Hundreds | Tens | Ones |
| | | 1, | 3 | 9 | 2, | 0 | 0 | 0 |

A special set of rules which gives the order in which calculations are done in an expression

# Going to London, England

Image Credits: (bg) ©Digital Vision/Getty Images, (b) ©Corbis

© Houghton Mifflin Harcourt Publishing Company

**Word Box**

base

Distributive
   Property

evaluate

exponent

inverse operations

numerical
   expression

order of operations

period

For 2 to 4 players

## Materials

- playing pieces: 3 of each color per player: red, blue, green, and yellow
- 1 number cube

## How to Play

1. Put your 3 playing pieces in the START circle of the same color.
2. To get a playing piece out of START, you must toss a 6.
    - If you toss a 6, move 1 of your playing pieces to the same-colored circle on the path.
    - If you do not toss a 6, wait until your next turn.
3. Once you have a playing piece on the path, toss the number cube to take a turn. Move the playing piece that many tan spaces. You must get all three of your playing pieces on the path.
4. If you land on a space with a question, answer it. If you are correct, move ahead 1 space.
5. To reach FINISH move your playing pieces along the path that is the same color as your playing piece. The first player to get all three playing pieces on FINISH wins.

**START**

Is this solution correct? Why or why not?
$36 - (8 \times 2) = 56$.

Use the order of operations to evaluate the expresion:
$6 + [(12 - 3) + (11 - 8)]$.

How can you write this number in two other forms:
$(8 \times 1000) + (9 \times 100) + (9 \times 1)$

What is the base in $10^3$?

Name two inverse operations.

What is an exponent?

**FINISH**

Write an expression: 48 cards are divided evenly among 6 friends.

**START**

Fill in the blanks: $7 \times 52 = (7 \times 50) + (\_\_ \times \_\_)$?

**START**

Fill in the blank:
If 108 ÷ 9 = 12,
then 9 × 12 = _____.

In the order of
operations, which comes
first: adding or dividing?

What does
the
Distributive
Property
say?

Use the
Distributive
Property
to rewrite
4 × 39.

**FINISH**

Fill in the
missing
exponent:
$10{,}000 = 10{\_}$

What does
it mean to
evaluate an
expression?

Write an expression:
Kim has 12 pencils.
She gives 10 to classmates.

Explain how to evaluate the
expression $(7 - 3) \times 6$.

**START**

Image Credits: (bg) ©c/Fotolia; (tl), (tr) ©Stockdisc/Getty Images; (bl) ©Markus Gann/
Shutterstock; (br) ©Thinkstock Images/Jupiterimages/Getty Images

# The Write Way

### Reflect

**Choose one idea. Write about it.**

- Explain how to use the Distributive Property to complete this equation.

  6 × (40 + 5) = (6 × _____) + (_____ × _____)

- Use the words *base* and *exponent* to tell how to rewrite this expression in exponent form.

  **10 × 10 × 10 × 10**

- Write two sentences to match this numerical expression: 7 × $3.

- Which solution uses the order of operations correctly? Explain how you know.

  **Solution A: 8 × 1 + 3 × 2 = 8 × 4 × 2 = 64**

  **Solution B: 8 × 1 + 3 × 2 = 8 + 6 = 14**

# Place Value and Patterns

**Essential Question** How can you describe the relationship between two place-value positions?

 **Common Core** **Number and Operations in Base Ten—5.NBT.A.1**
**MATHEMATICAL PRACTICES**
**MP2, MP5, MP7**

## Investigate

**Materials** ■ base-ten blocks

You can use base-ten blocks to understand the relationships among place-value positions. Use a large cube for 1,000, a flat for 100, a long for 10, and a small cube for 1.

| Number | 1,000 | 100 | 10 | 1 |
|---|---|---|---|---|
| Model | | | | |
| Description | large cube | flat | long | small cube |

**Complete the comparisons below to describe the relationship from one place-value position to the next place-value position.**

**A.** • Look at the long and compare it to the small cube.

The long is _____ times as much as the small cube.

• Look at the flat and compare it to the long.

The flat is _____ times as much as the long.

• Look at the large cube and compare it to the flat.

The large cube is _____ times as much as the flat.

**B.** • Look at the flat and compare it to the large cube.

The flat is _____ of the large cube.

• Look at the long and compare it to the flat.

The long is _____ of the flat.

• Look at the small cube and compare it to the long.

The small cube is _____ of the long.

**Math Talk** **MATHEMATICAL PRACTICES** ⑤

**Use Tools** How many times as much is the flat compared to the small cube? the large cube to the small cube? Explain.

1. **MATHEMATICAL PRACTICE 7** **Look for a Pattern** Describe the pattern you see when you move from a lesser place-value position to the next greater place-value position.

_____

_____

_____

2. **MATHEMATICAL PRACTICE 7** **Look for a Pattern** Describe the pattern you see when you move from a greater place-value position to the next lesser place-value position.

_____

_____

_____

## Make Connections

You can use your understanding of place-value patterns and a place-value chart to write numbers that are 10 times as much as or $\frac{1}{10}$ of any given number.

| Hundred Thousands | Ten Thousands | One Thousands | Hundreds | Tens | Ones |
|---|---|---|---|---|---|
| | | | 3 | 0 | 0 |
| | | ? | 300 | ? | |

10 times as much as      $\frac{1}{10}$ of

_____ is 10 times as much as 300.

_____ is $\frac{1}{10}$ of 300.

**Use the steps below to complete the table.**

**STEP 1** Write the given number in a place-value chart.

**STEP 2** Use the place-value chart to write a number that is 10 times as much as the given number.

**STEP 3** Use the place-value chart to write a number that is $\frac{1}{10}$ of the given number.

| Number | 10 times as much as | $\frac{1}{10}$ of |
|---|---|---|
| 10 | | |
| 70 | | |
| 9,000 | | |

Name _____

**Complete the sentence.**

1. 500 is 10 times as much as _____.

2. 20,000 is $\frac{1}{10}$ of _____.

3. 900 is $\frac{1}{10}$ of _____.

4. 600 is 10 times as much as _____.

## On Your Own

**Use place-value patterns to complete the table.**

| Number | 10 times as much as | $\frac{1}{10}$ of |
|---|---|---|
| 5. 10 | | |
| 6. 3,000 | | |
| 7. 800 | | |
| 8. 50 | | |

| Number | 10 times as much as | $\frac{1}{10}$ of |
|---|---|---|
| 9. 500 | | |
| 10. 90 | | |
| 11. 6,000 | | |
| 12. 200 | | |

**THINKSMARTER** **Complete the sentence with 100 or 1,000.**

13. 200 is _____ times as much as 2.

14. 4,000 is _____ times as much as 4.

15. 700,000 is _____ times as much as 700.

16. 600 is _____ times as much as 6.

## Problem Solving • Applications

17. **WRITE** *Math* Explain how you can use place-value patterns to describe how 50 and 5,000 compare.

_____

_____

18. **MATHEMATICAL PRACTICE ②** **Use Reasoning** 30,000 is _____ times as much as 30.

So, _____ is 10 times as much as 3,000.

**THINK SMARTER** **Sense or Nonsense?**

19. Mark and Robyn used base-ten blocks to show that 200 is 100 times as much as 2. Whose model makes sense? Whose model is nonsense? Explain your reasoning.

**Mark's Work**

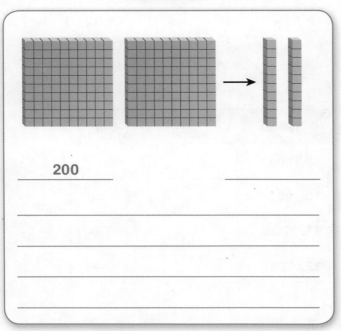

_____200_____ _____

_____

_____

_____

**Robyn's Work**

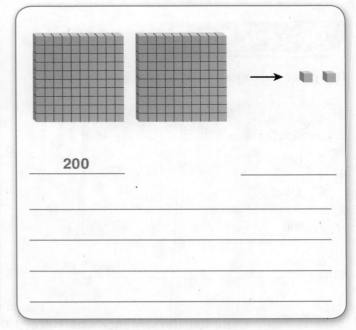

_____200_____ _____

_____

_____

_____

20. **GO DEEPER** Explain how you would help Mark understand why he should have used small cubes instead of longs.

_____

_____

_____

21. **THINK SMARTER** For 21a–21c, choose True or False for each sentence.

21a. 600 is $\frac{1}{10}$ of 6,000. ○ True ○ False

21b. 67 is $\frac{1}{10}$ of 6,700. ○ True ○ False

21c. 1,400 is 10 times as much as 140. ○ True ○ False

## Place Value and Patterns

Common Core  **COMMON CORE STANDARD—5.NBT.A.1**
*Understand the place value system.*

**Complete the sentence.**

1. 40,000 is 10 times as much as _____ 4,000 _____.

2. 90 is $\frac{1}{10}$ of _____.

3. 800 is 10 times as much as _____.

4. 5,000 is $\frac{1}{10}$ of _____.

**Use place-value patterns to complete the table.**

| Number | 10 times as much as | $\frac{1}{10}$ of |
|--------|---------------------|-------------------|
| 5. 100 | | |
| 6. 7,000 | | |
| 7. 80 | | |

| Number | 10 times as much as | $\frac{1}{10}$ of |
|--------|---------------------|-------------------|
| 8. 2,000 | | |
| 9. 400 | | |
| 10. 60 | | |

## Problem Solving (Real World)

11. The Eatery Restaurant has 200 tables. On a recent evening, there were reservations for $\frac{1}{10}$ of the tables. How many tables were reserved?

_____

12. Mr. Wilson has $3,000 in his bank account. Ms. Nelson has 10 times as much money in her bank account as Mr. Wilson has in his bank account. How much money does Ms. Nelson have in her bank account?

_____

13. **WRITE** ▸*Math* Write a number that has four digits with the same number in all places, such as 4,444. Circle the digit with the greatest value. Underline the digit with the least value. Explain.

_____

_____

_____

## Lesson Check (5.NBT.A.1)

**1.** What is 10 times as much as 700?

**2.** What is $\frac{1}{10}$ of 3,000?

## Spiral Review (Reviews 4.OA.A.3, 4.NBT.A.2, 4.NBT.B.5, 4.MD.A.3)

**3.** Risa is sewing a ribbon around the sides of a square blanket. Each side of the blanket is 72 inches long. How many inches of ribbon will Risa need?

**4.** What is the value of $n$?

$$9 \times 27 + 2 \times 31 - 28 = n$$

**5.** What is the best estimate for the product of 289 and 7?

**6.** Arrange the following numbers in order from greatest to least: 7,361; 7,136; 7,613

FOR MORE PRACTICE
GO TO THE
**Personal Math Trainer**

Name _____

# Place Value of Whole Numbers

**Essential Question** How do you read, write, and represent whole numbers through hundred millions?

 **Common Core** **Number and Operations in Base Ten—5.NBT.A.1**
MATHEMATICAL PRACTICES
MP1, MP2, MP7

 **Unlock the Problem** Real World

The diameter of the sun is 1,392,000 kilometers. To understand this distance, you need to understand the place value of each digit in 1,392,000.

A place-value chart contains periods. In numbers a **period** is a group of three digits separated by commas in a multidigit number. The millions period is left of the thousands period. One million is 1,000 thousands and is written as 1,000,000.

**Periods**

| MILLIONS | | | THOUSANDS | | | ONES | | |
|---|---|---|---|---|---|---|---|---|
| Hundreds | Tens | Ones | Hundreds | Tens | Ones | Hundreds | Tens | Ones |
| | | 1, | 3 | 9 | 2, | 0 | 0 | 0 |
| | | 1 × 1,000,000 | 3 × 100,000 | 9 × 10,000 | 2 × 1,000 | 0 × 100 | 0 × 10 | 0 × 1 |
| | | 1,000,000 | 300,000 | 90,000 | 2,000 | 0 | 0 | 0 |

The place value of the digit 1 in 1,392,000 is millions. The value of 1 in 1,392,000 is 1 × 1,000,000 = 1,000,000.

**Standard Form:** 1,392,000
**Word Form:** one million, three hundred ninety-two thousand
**Expanded Form:**
$(1 × 1,000,000) + (3 × 100,000) + (9 × 10,000) + (2 × 1,000)$

**Math Idea**

When writing a number in expanded form, if no digits appear in a place value, it is not necessary to include them in the expression.

---

**Try This!** Use place value to read and write numbers.

**Standard Form:** 582,030

**Word Form:** five hundred eighty-two _____, _____

**Expanded Form:** $(5 × 100,000) + ($ _____ $×$ _____ $) + (2 × 1,000) + ($ _____ $×$ _____ $)$

---

• The average distance from Jupiter to the sun is four hundred eighty-three million, six hundred thousand miles. Write the

number that shows this distance in miles. _____

# Place-Value Patterns

Canada's land area is about 4,000,000 square miles.
Iceland has a land area of about 40,000 square miles.
Compare the two areas.

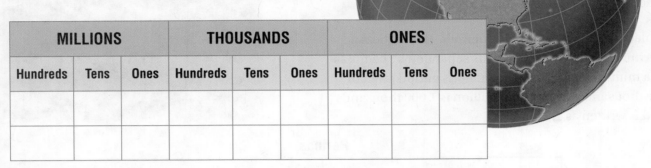

## 🔓 Example 1 Use a place-value chart.

**STEP 1**   Write the numbers in a place-value chart.

| MILLIONS | | | THOUSANDS | | | ONES | | |
|---|---|---|---|---|---|---|---|---|
| Hundreds | Tens | Ones | Hundreds | Tens | Ones | Hundreds | Tens | Ones |
| | | | | | | | | |
| | | | | | | | | |

**STEP 2**

Count the number of whole number place-value positions.

4,000,000 has _____ more whole number places than 40,000.

**Think:** 2 more places is 10 × 10, or 100.

4,000,000 is _____ times as much as 40,000.

So, Canada's estimated land area is _____ times as much as Iceland's estimated land area.

You can use place-value patterns to rename a number.

## 🔓 Example 2 Use place-value patterns.

Rename 40,000 using other place values.

| 40,000 | 4 ten thousands | 4 × 10,000 |
|---|---|---|
| 40,000 | _____ thousands | _____ × 1,000 |
| 40,000 | _____ | _____ |

Name _____

## Share and Show  MATH BOARD

**1.** Complete the place-value chart to find the value of each digit.

| MILLIONS | | | THOUSANDS | | | ONES | | |
|---|---|---|---|---|---|---|---|---|
| Hundreds | Tens | Ones | Hundreds | Tens | Ones | Hundreds | Tens | Ones |
| | | 7, | 3 | 3 | 3, | 8 | 2 | 0 |
| | | 7 × 1,000,000 | 3 × _____ | 3 × 10,000 | _____ × 1,000 | 8 × 100 | _____ | 0 × 1 |
| | | _____ | _____ | 30,000 | 3,000 | _____ | 20 | 0 |

**Write the value of the underlined digit.**

**2.** 1,57<u>4</u>,833

_____

**3.** 598,<u>1</u>02

_____

✓ **4.** 7,0<u>9</u>3,455

_____

**5.** <u>3</u>01,256,878

_____

**Write the number in two other forms.**

**6.** (8 × 100,000) + (4 × 1,000) + (6 × 1)

_____

_____

✓ **7.** seven million, twenty thousand, thirty-two

_____

_____

## On Your Own

**Write the value of the underlined digit.**

**8.** 8<u>4</u>9,567,043

_____

**9.** 9,<u>4</u>22,850

_____

**10.** <u>9</u>6,283

_____

**11.** <u>4</u>98,354,021

_____

**Write the number in two other forms.**

**12.** 345,000

_____

_____

_____

**13.** 119,000,003

_____

_____

_____

**14.** GoDEEPER  Consider the numbers 4,205,176 and 4,008.
What is the difference in the values of the digit 4 in each number?

_____

# Problem Solving • Applications

**Use the table for 15–16.**

| Average Distance from the Sun (in thousands of km) | | | |
|---|---|---|---|
| Mercury | 57,910 | Jupiter | 778,400 |
| Venus | 108,200 | Saturn | 1,427,000 |
| Earth | 149,600 | Uranus | 2,871,000 |
| Mars | 227,900 | Neptune | 4,498,000 |

**15.** Which planet is about 10 times as far as Earth is from the Sun?

_____

**16.** (MATHEMATICAL PRACTICE ❶) **Analyze Relationships** Which planet is about $\frac{1}{10}$ of the distance Uranus is from the Sun?

_____

**17.** THINKSMARTER **What's the Error?** Matt wrote the number four million, three hundred five thousand, seven hundred sixty-two as 4,350,762. Describe and correct his error.

_____

_____

WRITE ▸ Math • **Show Your Work**

**18.** GO DEEPER Explain how you know that the values of the digit 5 in the numbers 150,000 and 100,500 are not the same.

_____

_____

_____

_____

**19.** THINKSMARTER Select other ways to write 400,562. Mark all that apply.

(**A**) $(4 \times 100,000) + (50 \times 100) + (6 \times 10) + (2 \times 1)$

(**B**) four hundred thousand, five hundred sixty-two

(**C**) $(4 \times 100,000) + (5 \times 100) + (6 \times 10) + (2 \times 1)$

(**D**) four hundred, five hundred sixty-two

**Common Core** **COMMON CORE STANDARD—5.NBT.A.1**
*Understand the place value system.*

**Write the value of the underlined digit.**

**1.** 5,1<u>6</u>5,874

_____60,000_____

**2.** 2<u>8</u>1,480,100

_____

**3.** 7,<u>2</u>70

_____

**4.** 89,1<u>7</u>0,326

_____

**5.** <u>7</u>,050,423

_____

**6.** 6<u>4</u>6,950

_____

**7.** 37,<u>1</u>23,745

_____

**8.** <u>3</u>15,421,732

_____

**Write the number in two other forms.**

**9.** 15,409

_____

_____

**10.** 100,203

_____

_____

**Problem Solving** *Real World*

**11.** The U.S. Census Bureau has a population clock on the Internet. On a recent day, the United States population was listed as 310,763,136. Write this number in word form.

_____

_____

_____

**12.** In 2008, the population of 10- to 14-year-olds in the United States was 20,484,163. Write this number in expanded form.

_____

_____

_____

**13.** **WRITE** *Math* Write *Standard Form, Expanded Form,* and *Word Form* at the top of the page. Write five numbers that are at least 8 digits long under Standard Form. Write the expanded form and the word form for each number under the appropriate heading.

## Lesson Check (5.NBT.A.1)

**1.** A movie cost $3,254,107 to produce. What digit is in the hundred thousands place?

_____

**2.** What is the standard form of two hundred ten million, sixty-four thousand, fifty?

_____

## Spiral Review (Reviews 4.OA.C.5, 4.NBT.B.6, 4.G.A.2, 4.G.A.3)

**3.** If the pattern below continues, what number likely comes next?

9, 12, 15, 18, 21, __?__

_____

**4.** Find the quotient and remainder for 52 ÷ 8.

_____

**5.** How many pairs of parallel sides does the trapezoid below have?

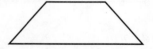

_____

**6.** How many lines of symmetry does the figure below appear to have?

_____

FOR MORE PRACTICE
GO TO THE
**Personal Math Trainer**

# Properties

**Essential Question** How can you use properties of operations to solve problems?

You can use the properties of operations to help you evaluate numerical expressions more easily.

 **Common Core** Operations and Algebraic Thinking—5.OA.A.1
**MATHEMATICAL PRACTICES**
MP1, MP2, MP8

## Properties of Addition

| | |
|---|---|
| **Commutative Property of Addition** <br> If the order of addends changes, the sum stays the same. | $12 + 7 = 7 + 12$ |
| **Associative Property of Addition** <br> If the grouping of addends changes, the sum stays the same. | $5 + (8 + 14) = (5 + 8) + 14$ |
| **Identity Property of Addition** <br> The sum of any number and 0 is that number. | $13 + 0 = 13$ |

## Properties of Multiplication

| | |
|---|---|
| **Commutative Property of Multiplication** <br> If the order of factors changes, the product stays the same. | $4 \times 9 = 9 \times 4$ |
| **Associative Property of Multiplication** <br> If the grouping of factors changes, the product stays the same. | $11 \times (3 \times 6) = (11 \times 3) \times 6$ |
| **Identity Property of Multiplication** <br> The product of any number and 1 is that number. | $4 \times 1 = 4$ |

 **Unlock the Problem**  Real World

The table shows the number of bones in several parts of the human body. What is the total number of bones in the ribs, the skull, and the spine?

To find the sum of addends using mental math, you can use the Commutative and Associative Properties.

| Part | Number of Bones |
|---|---|
| Ankle | 7 |
| Ribs | 24 |
| Skull | 28 |
| Spine | 26 |

🔑 **Use properties to find 24 + 28 + 26.**

$24 + 28 + 26 = 28 + $ _____ $ + 26$      Use the _____ Property to reorder the addends.

$= 28 + (24 + $ _____ $)$      Use the _____ Property to group the addends.

$= 28 + $ _____      Use mental math to add.

$= $ _____

So, there are _____ bones in the ribs, the skull, and the spine.

 **Math Talk**

**MATHEMATICAL PRACTICES ⑧**

**Generalize** Explain why grouping 24 and 26 makes the problem easier to solve.

## Distributive Property

Multiplying a sum by a number is the same as multiplying each addend by the number and then adding the products.

$5 \times (7 + 9) = (5 \times 7) + (5 \times 9)$

The Distributive Property can also be used with multiplication and subtraction. For example, $2 \times (10 - 8) = (2 \times 10) - (2 \times 8)$.

---

### 🔒 Example 1 Use the Distributive Property to find the product.

#### One Way Use addition.

$8 \times 59 = 8 \times ($_____$+ 9)$       Use a multiple of 10 to write 59 as a sum.

$= ($_____$\times 50) + (8 \times$_____$)$       Use the Distributive Property.

$=$_____$+$_____       Use mental math to multiply.

$=$_____       Use mental math to add.

#### Another Way Use subtraction.

$8 \times 59 = 8 \times ($_____$- 1)$       Use a multiple of 10 to write 59 as a difference.

$= ($_____$\times 60) - (8 \times$_____$)$       Use the Distributive Property.

$=$_____$-$_____       Use mental math to multiply.

$=$_____       Use mental math to subtract.

---

### 🔒 Example 2 Complete the equation, and tell which property you used.

**A** $23 \times$_____$= 23$

**Think:** A number times 1 is equal to itself.

Property: _____

_____

**B** $47 \times 15 = 15 \times$_____

**Think:** Changing the order of factors does not change the product.

Property: _____

_____

**MATHEMATICAL PRACTICES** ❶

**Describe** how to use the Distributive Property to find the product $3 \times 299$.

Name _____

**1.** Use properties to find $4 \times 23 \times 25$.

$23 \times$ _____ $\times 25$          _____ Property of Multiplication

$23 \times ($ _____ $\times$ _____ $)$          _____ Property of Multiplication

$23 \times$ _____

_____

**Use properties to find the sum or product.**

**2.** $89 + 27 + 11$

_____

**3.** $9 \times 52$

_____

☑ **4.** $107 + 0 + 39 + 13$

_____

**Complete the equation, and tell which property you used.**

**5.** $9 \times (30 + 7) = (9 \times$ _____ $) + (9 \times 7)$

_____

☑ **6.** $0 +$ _____ $= 47$

_____

**Math Talk**

**MATHEMATICAL PRACTICES ①**

Describe how you can use properties to solve problems more easily.

## On Your Own

**Practice: Copy and Solve** Use properties to find the sum or product.

**7.** $3 \times 78$

**8.** $4 \times 60 \times 5$

**9.** $21 + 25 + 39 + 5$

**Complete the equation, and tell which property you used.**

**10.** $11 + (19 + 6) = (11 +$ _____ $) + 6$

_____

**11.** $25 + 14 =$ _____ $+ 25$

_____

**12.** **MATHEMATICAL PRACTICE ③ Apply** Show how you can use the Distributive Property to rewrite and find $(32 \times 6) + (32 \times 4)$.

_____

## Problem Solving • Applications

**13.** GO DEEPER   Three friends' meals at a restaurant cost $13, $14, and $11. Use parentheses to write two different expressions to show how much the friends spent in all. Which property does your pair of expressions demonstrate?

_____

_____

**14.** MATHEMATICAL PRACTICE ② **Use Reasoning**  Jacob is designing an aquarium for a doctor's office. He plans to buy 6 red blond guppies, 1 blue neon guppy, and 1 yellow guppy. The table shows the price list for the guppies. How much will the guppies for the aquarium cost?

_____

**15.** Sylvia bought 8 tickets to a concert. Each ticket costs $18. To find the total cost in dollars, she added the product $8 \times 10$ to the product $8 \times 8$, for a total of 144. Which property did Sylvia use?

_____

| Fancy Guppy Prices | |
|---|---|
| Blue neon | $11 |
| Red blond | $22 |
| Sunrise | $18 |
| Yellow | $19 |

WRITE ▸ *Math* • **Show Your Work**

**16.** THINK SMARTER   **Sense or Nonsense?** Julie wrote $(15 - 6) - 3 = 15 - (6 - 3)$. Is Julie's equation sense or nonsense? Do you think the Associative Property works for subtraction? Explain.

_____

_____

**17.** THINK SMARTER   Find the property that each equation shows.

$14 \times (4 \times 9) = (14 \times 4) \times 9$ •

$1 \times 3 = 3 \times 1$ •

$7 \times 3 = 3 \times 7$ •

 • Commutative Property of Multiplication

 • Associative Property of Multiplication

 • Identity Property of Multiplication

Name _____

## Properties

**COMMON CORE STANDARD—5.OA.A.1**
*Perform operations with multi-digit whole numbers and with decimals to hundredths.*

**Use properties to find the sum or product.**

**1.** $6 \times 89$

$6 \times (90 - 1)$

$(6 \times 90) - (6 \times 1)$

$540 - 6$

_____ 534

**2.** $93 + (68 + 7)$

_____

**3.** $5 \times 23 \times 2$

_____

**4.** $8 \times 51$

_____

**5.** $34 + 0 + 18 + 26$

_____

**6.** $6 \times 107$

_____

**Complete the equation, and tell which property you used.**

**7.** $(3 \times 10) \times 8 = $ _____ $ \times (10 \times 8)$

_____

_____

**8.** $16 + 31 = 31 + $ _____

_____

_____

## Problem Solving · Real World

**9.** The Metro Theater has 20 rows of seats with 18 seats in each row. Tickets cost $5. The theater's income in dollars if all seats are sold is $(20 \times 18) \times 5$. Use properties to find the total income.

_____

**10.** The numbers of students in the four sixth-grade classes at Northside School are 26, 19, 34, and 21. Use properties to find the total number of students in the four classes.

_____

**11.** **WRITE** ▸*Math* Explain how you could mentally find $8 \times 45$ by using the Distributive Property.

_____

_____

## Lesson Check (5.OA.A.1)

**1.** To find 19 + (11 + 37), Lennie added 19 and 11. Then he added 37 to the sum. What property did he use?

_____

_____

**2.** Marla did 65 sit-ups each day for one week. Use the Distributive Property to show an expression you can use to find the total number of sit-ups Marla did during the week.

_____

_____

## Spiral Review (Reviews 4.OA.B.4, 4.NBT.B.5, 4.NBT.B.6, 5.NBT.A.1)

**3.** The average sunflower has 34 petals. What is the best estimate of the total number of petals on 57 sunflowers?

_____

**4.** A golden eagle flies a distance of 290 miles in 5 days. If the eagle flies the same distance each day of its journey, how far does the eagle fly per day?

_____

**5.** What is the value of the underlined digit in the following number?

2,9<u>8</u>3,785

_____

**6.** What best describes the number 5? Write *prime, composite, neither prime nor composite, or both prime and composite.*

_____

**FOR MORE PRACTICE GO TO THE Personal Math Trainer**

Name _____

# Powers of 10 and Exponents

**Essential Question** How can you use an exponent to show powers of 10?

**Common Core** **Number and Operations in Base Ten—5.NBT.A.2**
**MATHEMATICAL PRACTICES**
**MP2, MP7, MP8**

## 🔑 Unlock the Problem

Expressions with repeated factors, such as $10 \times 10 \times 10$, can be written by using a base with an exponent. The **base** is the number that is used as the repeated factor. The **exponent** is the number that tells how many times the base is used as a factor.

$$10 \times 10 \times 10 = 10^3 = 1{,}000$$

3 factors    base      exponent

**Word form:** the third power of ten

**Exponent form:** $10^3$

## 🔒 Activity Use base-ten blocks.

**Materials** ■ base-ten blocks

What is $10 \times 1{,}000$ written with an exponent?

     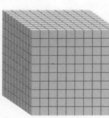

| 1 one | 10 ones | 100 ones | 1,000 ones |
|---|---|---|---|
| 1 | $1 \times 10$ | $1 \times 10 \times 10$ | $1 \times 10 \times 10 \times 10$ |
| $10^0$ | $10^1$ | $10^2$ | $10^3$ |

- How many ones are in 1? _____

- How many ones are in 10? _____

- How many tens are in 100? _____

 **Think:** 10 groups of 10 or $10 \times 10$

- How many hundreds are in 1,000? _____

 **Think:** 10 groups of 100 or $10 \times (10 \times 10)$

- How many thousands are in 10,000? _____

In the box at the right, draw a quick picture to show 10,000.

So, $10 \times 1{,}000$ is 10____.

▲ Use ⊤ for 1,000.

10,000 ones
$1 \times 10 \times 10 \times 10 \times 10$

10

# ① Example Multiply a whole number by a power of ten.

Hummingbirds beat their wings very fast. The smaller the hummingbird is, the faster its wings beat. The average hummingbird beats its wings about $3 \times 10^3$ times a minute. How many times a minute is that, written as a whole number?

Multiply 3 by powers of ten. Look for a pattern.

$3 \times 10^0 = 3 \times 1 =$ _____

$3 \times 10^1 = 3 \times 10 =$ _____

$3 \times 10^2 = 3 \times 10 \times 10 =$ _____

$3 \times 10^3 = 3 \times 10 \times 10 \times 10 =$ _____

So, the average hummingbird beats its wings about _____ times a minute.

**Math Talk** **MATHEMATICAL PRACTICES ⑧**
Generalize Explain how using an exponent simplifies an expression.

- **MATHEMATICAL PRACTICE ⑦ Look for a Pattern** What pattern do you see?

_____

_____

## Share and Show MATH BOARD

**Write in exponent form and word form.**

1. $10 \times 10$

   Exponent form: _____

   Word form: _____

   _____

✓ 2. $10 \times 10 \times 10 \times 10$

   Exponent form: _____

   Word form: _____

   _____

**Find the value.**

3. $10^2$

✓ 4. $4 \times 10^2$

5. $7 \times 10^3$

Name _____

**Write in exponent form and word form.**

**6.** $10 \times 10 \times 10$

exponent form: _____

word form: _____

_____

**7.** $10 \times 10 \times 10 \times 10 \times 10$

exponent form: _____

word form: _____

_____

**Find the value.**

**8.** $10^4$

_____

**9.** $2 \times 10^3$

_____

**10.** $6 \times 10^4$

_____

**GO DEEPER** **Complete the pattern.**

**11.** $12 \times 10^0 = 12 \times 1 =$ _____

$12 \times 10^1 = 12 \times 10 =$ _____

$12 \times 10^2 = 12 \times 100 =$ _____

$12 \times 10^3 = 12 \times 1{,}000 =$ _____

$12 \times 10^4 = 12 \times 10{,}000 =$ _____

**12.** **MATHEMATICAL PRACTICE ②** **Reason Abstractly** $10^3 = 10 \times 10^n$
What is the value of $n$?

**Think:** $10^3 = 10 \times$ _____ $\times$ _____,

or $10 \times$ _____

The value of $n$ is _____.

**13.** **WRITE** ▸ *Math* Explain how to write 50,000 using exponents.

_____

_____

**14.** **GO DEEPER** One year, Mr. James travels $9 \times 10^3$ miles for his job.
The next year he traveled $1 \times 10^4$ miles. How many more miles
did he travel the second year than he did the first year. Explain.

_____

_____

## Unlock the Problem Real World

**15.** **THINK SMARTER** Lake Superior is the largest of the Great Lakes. It covers a surface area of about 30,000 square miles. How can you show the estimated area of Lake Superior as a whole number multiplied by a power of ten?

a. What are you asked to find?

_____

b. How can you use a pattern to find the answer?

_____

c. Write a pattern using the whole number 3 and powers of ten.

$3 \times 10^0 = 3 \times 1 =$ _____

$3 \times 10^1 = 3 \times 10 =$ _____

$3 \times 10^2 =$ _____ $=$ _____

$3 \times 10^3 =$ _____ $=$ _____

$3 \times 10^4 =$ _____ $=$ _____

d. Complete the sentence.
The estimated area of Lake Superior is _____.

**16.** The Earth's diameter through the equator is about 8,000 miles. What is the Earth's estimated diameter written as a whole number multiplied by a power of ten?

_____

**17.** **THINK SMARTER** Yolanda says $10^5$ is the same as 50 because $10 \times 5$ equals 50. What was Yolanda's mistake?

_____

_____

## Powers of 10 and Exponents

Common Core **COMMON CORE STANDARD—5.NBT.A.2**
*Understand the place value system.*

**Write in exponent form and word form.**

**1.** $10 \times 10 \times 10$

**2.** $10 \times 10$

**3.** $10 \times 10 \times 10 \times 10$

exponent form: ____$10^3$____

exponent form: _____

exponent form: _____

word form: __the third power__

word form: _____

word form: _____

of ten _____

_____

_____

_____

_____

_____

**Find the value.**

**4.** $10^3$

**5.** $4 \times 10^2$

**6.** $7 \times 10^3$

**7.** $8 \times 10^0$

## Problem Solving · Real World

**8.** The moon is about 240,000 miles from Earth. What is this distance written as a whole number multiplied by a power of ten?

_____

**9.** The sun is about $93 \times 10^6$ miles from Earth. What is this distance written as a whole number?

_____

**10.** **WRITE** ▸ *Math* Consider $7 \times 10^3$. Write a pattern to find the value of the expression.

_____

_____

## Lesson Check (5.NBT.A.2)

1. Write the expression that shows "3 times the sixth power of 10."

2. Gary mails $10^3$ flyers to clients in one week. How many flyers does Gary mail?

## Spiral Review (Reviews 4.NBT.B.5, 4.NBT.B.6)

3. Harley is loading 625 bags of cement onto small pallets. Each pallet holds 5 bags. How many pallets will Harley need?

4. Marylou buys a package of 500 jewels to decorate 4 different pairs of jeans. She uses the same number of jewels on each pair of jeans. How many jewels will she use for each pair of jeans?

5. Manny buys 4 boxes of straws for his restaurant. There are 500 straws in each box. How many straws does he buy?

6. Cammie goes to the gym to exercise 4 times per week. Altogether, how many times does she go to the gym in 10 weeks?

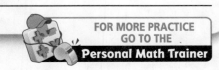

FOR MORE PRACTICE
GO TO THE
Personal Math Trainer

Name _____

# Multiplication Patterns

**Essential Question** How can you use a basic fact and a pattern to multiply by a 2-digit number?

**Common Core** **Number and Operations in Base Ten—5.NBT.A.2**
**MATHEMATICAL PRACTICES**
**MP2, MP3, MP8**

 ## Unlock the Problem  Real World

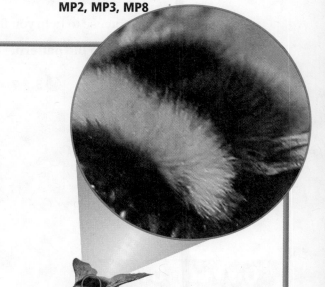

How close have you been to a bumblebee?

The actual length of a queen bumblebee is about 20 millimeters. The photograph shows part of a bee under a microscope, at 10 times its actual size. What would the length of the bee appear to be at a magnification of 300 times its actual size?

**Use a basic fact and a pattern.**

**Multiply.** $300 \times 20$

$3 \times 2 = 6$ ← basic fact

$30 \times 2 = (3 \times 2) \times 10^1 = 60$

$300 \times 2 = (3 \times 2) \times 10^2 = $ _____

$300 \times 20 = (3 \times 2) \times (100 \times 10) = 6 \times 10^3 = $ _____

So, the length of the bee would appear to be

about _____ millimeters.

**Math Talk**

**MATHEMATICAL PRACTICES ⑧**

Generalize What pattern do you see in the number sentences and the exponents?

- What would the length of the bee shown in the photograph appear to be if the microscope shows it at 10 times its actual size?

  _____

## Example Use mental math and a pattern.

**Multiply.** $50 \times 8{,}000$

$5 \times 8 = 40$ ← basic fact

$5 \times 80 = (5 \times 8) \times 10^1 = 400$

$5 \times 800 = (5 \times 8) \times 10^2 = $ _____

$50 \times 800 = (5 \times 8) \times (10 \times 100) = 40 \times 10^3 = $ _____

$50 \times 8{,}000 = (5 \times 8) \times (10 \times 1{,}000) = 40 \times 10^4 = $ _____

**Use mental math and a pattern to find the product.**

**1.** $30 \times 4{,}000 =$ _____

What basic fact can you use to help you find $30 \times 4{,}000$? _____

**Use mental math to complete the pattern.**

**2.** $1 \times 1 = 1$

$1 \times 10^1 =$ _____

$1 \times 10^2 =$ _____

$1 \times 10^3 =$ _____

**☑ 3.** $7 \times 8 = 56$

$(7 \times 8) \times 10^1 =$ _____

$(7 \times 8) \times 10^2 =$ _____

$(7 \times 8) \times 10^3 =$ _____

**☑ 4.** $6 \times 5 =$ _____

$(6 \times 5) \times$ _____ $= 300$

$(6 \times 5) \times$ _____ $= 3{,}000$

$(6 \times 5) \times$ _____ $= 30{,}000$

**Math Talk** — MATHEMATICAL PRACTICES ❸

Apply Tell how to find $50 \times 9{,}000$ by using a basic fact and pattern.

## On Your Own

**Use mental math to complete the pattern.**

**5.** $9 \times 5 = 45$

$(9 \times 5) \times 10^1 =$ _____

$(9 \times 5) \times 10^2 =$ _____

$(9 \times 5) \times 10^3 =$ _____

**6.** $3 \times 7 = 21$

$(3 \times 7) \times 10^1 =$ _____

$(3 \times 7) \times 10^2 =$ _____

$(3 \times 7) \times 10^3 =$ _____

**7.** $5 \times 4 =$ _____

$(5 \times 4) \times$ _____ $= 200$

$(5 \times 4) \times$ _____ $= 2{,}000$

$(5 \times 4) \times$ _____ $= 20{,}000$

**Use mental math and a pattern to find the product.**

**8.** $(6 \times 6) \times 10^1 =$ _____

**9.** $(7 \times 4) \times 10^3 =$ _____

**10.** $(9 \times 8) \times 10^2 =$ _____

**11.** $(4 \times 3) \times 10^2 =$ _____

**12.** $(2 \times 5) \times 10^3 =$ _____

**13.** $(2 \times 8) \times 10^2 =$ _____

**14.** $(6 \times 5) \times 10^3 =$ _____

**15.** $(8 \times 8) \times 10^4 =$ _____

**16.** $(7 \times 8) \times 10^4 =$ _____

**17.** _THINK SMARTER_ What does the product of any whole-number factor multiplied by 100 always have? Explain.

_____

_____

_____

Name _____

**Use mental math to complete the table.**

**18.** 1 roll = 50 dimes   **Think:** 50 dimes per roll × 20 rolls = (5 × 2) × (10 × 10)

| Rolls | 20 | 30 | 40 | 50 | 60 | 70 | 80 | 90 | 100 |
|-------|----|----|----|----|----|----|----|----|-----|
| Dimes | $10 \times 10^2$ | | | | | | | | |

**19.** 1 roll = 40 quarters   **Think:** 40 quarters per roll × 20 rolls = (4 × 2) × (10 × 10)

| Rolls | 20 | 30 | 40 | 50 | 60 | 70 | 80 | 90 | 100 |
|-------|----|----|----|----|----|----|----|----|-----|
| Quarters | $8 \times 10^2$ | | | | | | | | |

| × | 6 | 70 | 800 | 9,000 |
|---|---|----|-----|-------|
| **20.** 80 | | | $64 \times 10^3$ | |
| **21.** 90 | | | | $81 \times 10^4$ |

## Problem Solving • Applications

**Use the table for 22–24.**

**22.** **What if** you magnified the image of a cluster fly by $9 \times 10^3$? What would the length appear to be?

_____

**23.** **GO DEEPER** If you magnified the images of a fire ant by $4 \times 10^3$ and a tree hopper by $3 \times 10^3$, which insect would appear longer? How much longer?

_____

**24.** **MATHEMATICAL PRACTICE ②** **Reason Quantitatively** John wants to magnify the image of a fire ant and a crab spider so they appear to be the same length. How many times their actual sizes would he need to magnify each image?

_____

_____

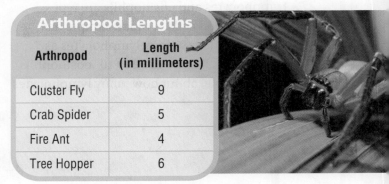

**Arthropod Lengths**

| Arthropod | Length (in millimeters) |
|-----------|-------------------------|
| Cluster Fly | 9 |
| Crab Spider | 5 |
| Fire Ant | 4 |
| Tree Hopper | 6 |

**WRITE** ▸*Math* • **Show Your Work**

## Connect to Health

### Blood Cells

Blood is necessary for all human life. It contains red blood cells and white blood cells that nourish and cleanse the body and platelets that stop bleeding. The average adult has about 5 liters of blood.

◄ Single red blood cell

▲ Platelet

White blood cell ►

**Use patterns and mental math to solve.**

**25.** GO DEEPER A human body has about 30 times as many platelets as white blood cells. A small sample of blood has $8 \times 10^3$ white blood cells. About how many platelets are in the sample?

**26.** Basophils and monocytes are types of white blood cells. A blood sample has about 5 times as many monocytes as basophils. If there are 60 basophils in the sample, about how many monocytes are there?

**27.** Lymphocytes and eosinophils are types of white blood cells. A blood sample has about 10 times as many lymphocytes as eosinophils. If there are $2 \times 10^2$ eosinophils in the sample, about how many lymphocytes are there?

**28.** THINK SMARTER An average person has $6 \times 10^2$ times as many red blood cells as white blood cells. A small sample of blood has $7 \times 10^3$ white blood cells. About how many red blood cells are in the sample?

**29.** THINK SMARTER Kyle says $20 \times 10^4$ is the same as 20,000. He reasoned that since he saw 4 as the exponent he should write 4 zeros in his answer. Is Kyle correct?

# Multiplication Patterns

Common Core    **COMMON CORE STANDARD—5.NBT.A.2**
*Understand the place value system.*

**Use mental math to complete the pattern.**

**1.** $8 \times 3 = 24$

$(8 \times 3) \times 10^1 =$ ___240___

$(8 \times 3) \times 10^2 =$ ___2,400___

$(8 \times 3) \times 10^3 =$ ___24,000___

**2.** $5 \times 6 =$ _____

$(5 \times 6) \times 10^1 =$ _____

$(5 \times 6) \times 10^2 =$ _____

$(5 \times 6) \times 10^3 =$ _____

**3.** $3 \times$ _____ $= 27$

$(3 \times 9) \times 10^1 =$ _____

$(3 \times 9) \times 10^2 =$ _____

$(3 \times 9) \times 10^3 =$ _____

**4.** _____ $\times 4 = 28$

$(7 \times 4) \times$ _____ $= 280$

$(7 \times 4) \times$ _____ $= 2,800$

$(7 \times 4) \times$ _____ $= 28,000$

**5.** $6 \times 8 =$ _____

$(6 \times 8) \times 10^2 =$ _____

$(6 \times 8) \times 10^3 =$ _____

$(6 \times 8) \times 10^4 =$ _____

**6.** _____ $\times 4 = 16$

$(4 \times 4) \times 10^2 =$ _____

$(4 \times 4) \times 10^3 =$ _____

$(4 \times 4) \times 10^4 =$ _____

**Use mental math and a pattern to find the product.**

**7.** $(2 \times 9) \times 10^2 =$ _____

**8.** $(8 \times 7) \times 10^2 =$ _____

**9.** $(3 \times 7) \times 10^3 =$ _____

**10.** $(5 \times 9) \times 10^4 =$ _____

**11.** $(4 \times 8) \times 10^4 =$ _____

**12.** $(8 \times 8) \times 10^3 =$ _____

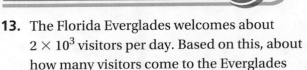

**Problem Solving** Real World

**13.** The Florida Everglades welcomes about $2 \times 10^3$ visitors per day. Based on this, about how many visitors come to the Everglades per week?

_____

_____

**14.** The average person loses about $8 \times 10^1$ strands of hair each day. About how many strands of hair would the average person lose in 9 days?

_____

_____

**15.** **WRITE** ▸*Math*  Do the products $40 \times 500$ and $40 \times 600$ have the same number of zeros? Explain.

_____

_____

## Lesson Check (5.NBT.A.2)

**1.** How many zeros are in the product $(6 \times 5) \times 10^3$?

**2.** Addison studies a tarantula that is 30 millimeters long. Suppose she uses a microscope to magnify the spider by $4 \times 10^2$. How long will the spider appear to be?

## Spiral Review (Reviews 4.OA.A.3, 4.NBT.B.5)

**3.** Hayden has 6 rolls of dimes. There are 50 dimes in each roll. How many dimes does he have altogether?

**4.** An adult ticket to the zoo costs $20 and a child's ticket costs $10. How much will it cost for Mr. and Mrs. Brown and their 4 children to get into the zoo?

**5.** At a museum, 100 posters are displayed in each of 4 rooms. Altogether, how many posters are displayed?

**6.** A store sells a gallon of milk for $3. A baker buys 30 gallons of milk for his bakery. How much will he have to pay?

**FOR MORE PRACTICE
GO TO THE
Personal Math Trainer**

Name _____

# ☑ Mid-Chapter Checkpoint

**Personal Math Trainer**
Online Assessment
and Intervention

## Vocabulary

**Choose the best term from the box.**

Vocabulary

base

exponent

period

1. A group of three digits separated by commas in a multidigit

   number is a _____. (p. 11)

2. An _____ is the number that tells how many times a base is
   used as a factor. (p. 23)

## Concepts and Skills

**Complete the sentence.** (5.NBT.A.1)

3. 7 is $\frac{1}{10}$ of _____.

4. 800 is 10 times as much as _____.

**Write the value of the underlined digit.** (5.NBT.A.1)

5. 6,5<u>8</u>1,678

6. 125,<u>6</u>34

7. 3<u>4</u>,634,803

8. 2,<u>7</u>64,835

_____   _____   _____   _____

**Complete the equation, and tell which property you used.** (5.OA.A.1)

9. $8 \times (14 + 7) =$ _____ $+ (8 \times 7)$

10. $7 + (8 + 12) =$ _____ $+ 12$

_____   _____

**Find the value.** (5.NBT.A.2)

11. $10^3$

12. $6 \times 10^2$

13. $4 \times 10^4$

_____   _____   _____

**Use mental math and a pattern to find the product.** (5.NBT.A.2)

14. $70 \times 300 =$ _____

15. $(3 \times 4) \times 10^3 =$ _____

**16.** DVDs are on sale for $24 each. Felipe writes the expression 4 × 24 to find the cost in dollars of buying 4 DVDs. How can you rewrite Felipe's expression using the Distributive Property? (5.OA.A.1)

_____

**17.** The Muffin Shop chain of bakeries sold 745,305 muffins last year. Write this number in expanded form. (5.NBT.A.1)

_____

_____

**18.** The soccer field at Mario's school has an area of 6,000 square meters. How can Mario show the area as a whole number multiplied by a power of ten? (5.NBT.A.2)

_____

**19.** Ms. Alonzo ordered 4,000 markers for her store. Only $\frac{1}{10}$ of them arrived. How many markers did she receive? (5.NBT.A.1)

_____

**20.** GO DEEPER  Mark wrote the highest score he made on his new video game as the product of 70 × 6,000. Use the Associative and Commutative Properties to show how Mark can calculate this product mentally. (5.NBT.A.2)

_____

_____

Name _____

# Multiply by 1-Digit Numbers

**Essential Question** How do you multiply by 1-digit numbers?

##  Unlock the Problem Real World

Each day an airline flies 9 commercial jets from New York to London, England. Each plane holds 293 passengers. If every seat is taken on all flights, how many passengers fly on this airline from New York to London in 1 day?

🔑 **Use place value and regrouping.**

**STEP 1** Estimate: $293 \times 9$

> Think: $300 \times 9 =$ _____

▲ The Queen's Guard protects Britain's Royal Family and their residences.

**STEP 2** Multiply the ones.

$$\begin{array}{r} \overset{2}{29}3 \\ \times\ 9 \\ \hline 7 \end{array}$$

$9 \times 3$ ones = _____ ones

Write the ones and the regrouped tens.

**Math Talk**

MATHEMATICAL PRACTICES ①

Describe how to record the 27 ones when you multiply 3 by 9 in Step 2.

**STEP 3** Multiply the tens.

$$\begin{array}{r} \overset{8}{\underset{2}{2}}93 \\ \times\ 9 \\ \hline 37 \end{array}$$

$9 \times 9$ tens = _____ tens

Add the regrouped tens.

_____ tens + 2 tens = _____ tens

Write the tens and the regrouped hundreds.

**STEP 4** Multiply the hundreds.

$$\begin{array}{r} \overset{8}{\underset{2}{2}}93 \\ \times\ 9 \\ \hline 2{,}637 \end{array}$$

$9 \times 2$ hundreds = _____ hundreds

Add the regrouped hundreds.

_____ hundreds + 8 hundreds = _____ hundreds

Write the hundreds.

So, in 1 day, _____ passengers fly from New York to London.

• **MATHEMATICAL PRACTICE ①** **Evaluate Reasonableness** How can you tell if your answer is reasonable? _____

_____

## 🔒 Example

A commercial airline makes several flights each week from New York to Paris, France. If the airline serves 1,978 meals on its flights each day, how many meals are served for the entire week?

To multiply a greater number by a 1-digit number, repeat the process of multiplying and regrouping until every place value is multiplied.

**STEP 1** Estimate. $1,978 \times 7$

   Think: $2,000 \times 7 =$ _____

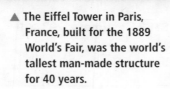

▲ The Eiffel Tower in Paris, France, built for the 1889 World's Fair, was the world's tallest man-made structure for 40 years.

**STEP 2** Multiply the ones.

$$\begin{array}{r} \overset{5}{1,97}8 \\ \times \quad 7 \\ \hline 6 \end{array}$$

$7 \times 8$ ones = _____ ones

Write the ones and the regrouped tens.

**STEP 3** Multiply the tens.

$$\begin{array}{r} \overset{55}{1,9}78 \\ \times \quad 7 \\ \hline 46 \end{array}$$

$7 \times 7$ tens = _____ tens

Add the regrouped tens.

_____ tens + 5 tens = _____ tens

Write the tens and the regrouped hundreds.

**STEP 4** Multiply the hundreds.

$$\begin{array}{r} \overset{6\ 55}{1,9}78 \\ \times \quad 7 \\ \hline 846 \end{array}$$

$7 \times 9$ hundreds = _____ hundreds

Add the regrouped hundreds.

_____ hundreds + 5 hundreds = _____ hundreds

Write the hundreds and the regrouped thousands.

**STEP 5** Multiply the thousands.

$$\begin{array}{r} \overset{6\ 55}{1,9}78 \\ \times \quad 7 \\ \hline 13,846 \end{array}$$

$7 \times 1$ thousand = _____ thousands

Add the regrouped thousands.

_____ thousands + 6 thousands = _____ thousands

Write the thousands. Compare your answer to the estimate to see if it is reasonable.

So, in 1 week, _____ meals are served on flights from New York to Paris.

Name _____

**Complete to find the product.**

**1.** 6 × 796        **Estimate:** 6 × _____ = _____

796    Multiply the ones
× 6    and regroup.

<sup>3</sup>
796    Multiply the
× 6    tens and add the
___    regrouped tens.
6      Regroup.

<sup>53</sup>
796    Multiply the
× 6    hundreds and add
___    the regrouped
76     hundreds.

**Estimate. Then find the product.**

**2.** Estimate: _____

608
× 8
____

**3.** Estimate: _____

556
× 4
____

**4.** Estimate: _____

1,925
× 7
_____

## On Your Own

**MATHEMATICAL PRACTICE ②** Use Reasoning **Algebra** Solve for the unknown numbers.

**5.**
396
× 6
____
2,3⬜6

**6.**
5,12⬜
× 8
____
⬜⬜16

**7.**
8,5⬜6
× 7
_____
60,03⬜

**Practice: Copy and Solve** Estimate. Then find the product.

**8.** 116 × 3        **9.** 338 × 4        **10.** 6 × 219        **11.** 7 × 456

**12.** **THINK SMARTER** A commercial airline makes a flight each day from New York to Paris, France. The aircraft seats 524 passengers and serves 2 meals to each passenger per flight. If all the seats are filled each flight, how many meals are served in one week?

_____

## Problem Solving • Applications

13. **THINK SMARTER** **What's the Error?** The Plattsville Glee Club is sending 8 of its members to a singing contest in Cincinnati, Ohio. The cost will be $588 per person. How much will it cost for the entire group of 8 students to attend?

Both Brian and Jermaine solve the problem. Brian says the answer is $40,704. Jermaine's answer is $4,604.

Estimate the cost. A reasonable estimate is _____.

Although Jermaine's answer seems reasonable, neither Brian nor Jermaine solved the problem correctly. Find the errors in Brian's and Jermaine's work. Then, solve the problem correctly.

| **Brian** | **Jermaine** | **Correct Answer** |
|:---:|:---:|:---:|
| 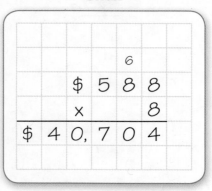 | 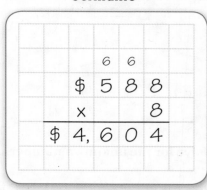 | 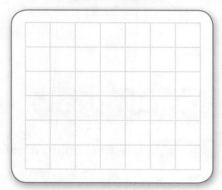 |

- **MATHEMATICAL PRACTICE ❸ Verify the Reasoning of Others** What error did Brian make? Explain.

  _____

  _____

- What error did Jermaine make? Explain.

  _____

  _____

14. **GO DEEPER** How could you predict that Jermaine's answer might be incorrect

using your estimate? _____

_____

_____

## Multiply by 1-Digit Numbers

Common Core

**COMMON CORE STANDARD—5.NBT.B.6**
*Perform operations with multi-digit whole numbers and with decimals to hundredths.*

**Estimate. Then find the product.**

**1.** Estimate: ___3,600___

$$
\begin{array}{r}
1\ 5 \\
416 \\
\times\ \ 9 \\
\hline
3,744
\end{array}
$$

**2.** Estimate: _____

$$
\begin{array}{r}
1,374 \\
\times\ \ \ \ 6 \\
\hline
\end{array}
$$

**3.** Estimate: _____

$$
\begin{array}{r}
726 \\
\times\ \ \ 5 \\
\hline
\end{array}
$$

**Estimate. Then find the product.**

**4.** $4 \times 979$

**5.** $503 \times 7$

**6.** $5 \times 4,257$

**7.** $6,018 \times 9$

**8.** $758 \times 6$

**9.** $3 \times 697$

**10.** $2,141 \times 8$

**11.** $7 \times 7,956$

## Problem Solving Real World

**12.** Mr. and Mrs. Dorsey and their three children are flying to Springfield. The cost of each ticket is $179. Estimate how much the tickets will cost. Then find the exact cost of the tickets.

_____

_____

_____

**13.** Ms. Tao flies roundtrip twice yearly between Jacksonville and Los Angeles on business. The distance between the two cities is 2,150 miles. Estimate the distance she flies for both trips. Then find the exact distance.

_____

_____

_____

**14.** **WRITE** ▸*Math* Show how to solve the problem $378 \times 6$ using place value with regrouping. Explain how you knew when to regroup.

_____

_____

## Lesson Check (5.NBT.B.5)

1. Mr. Nielson works 154 hours each month. He works 8 months each year. How many hours does Mr. Nielson work each year?

2. Sasha lives 1,493 miles from her grandmother. One year, Sasha's family made 4 round trips to visit her grandmother. How many miles did they travel in all?

## Spiral Review (Reviews 4.NBT.A.2, 4.NBT.A.3, 4.NF.C.6, 5.NBT.A.1)

3. Yuna missed 5 points out of 100 points on her math test. What decimal number represents the part of her math test that she answered correctly?

4. Which symbol makes the statement true? Write >, <, or =.

   602,163 ◯ 620,163

5. The number below represents the number of fans that attended Chicago Cubs baseball games in 2008. What is this number written in standard form?

   $(3 \times 1,000,000) + (3 \times 100,000) + (2 \times 100)$

6. A fair was attended by 755,082 people altogether. What is this number rounded to the nearest ten thousand?

FOR MORE PRACTICE
GO TO THE
Personal Math Trainer

# Multiply by Multi-Digit Numbers

**Essential Question** How do you multiply by multi-digit numbers?

 **Common Core** **Number and Operations in Base Ten—5.NBT.B.5**
**MATHEMATICAL PRACTICES**
**MP1, MP4, MP6**

 **Unlock the Problem** *Real World*

A tiger can eat as much as 40 pounds of food at a time but it may go for several days without eating anything. Suppose a Siberian tiger in the wild eats an average of 18 pounds of food per day. How much food will the tiger eat in 28 days if he eats that amount each day?

 **Use place value and regrouping.**

**STEP 1 Estimate:** 28 × 18

Think: 30 × 20 = _____

**STEP 2** Multiply by the ones.

          28
        × 18
     ▢          28 × 8 ones = _____ ones

**STEP 3** Multiply by the tens.

          28
        × 18
     ▢
     ▢          28 × 1 ten = _____ tens, or _____ ones

**STEP 4** Add the partial products.

          28
        × 18
     ▢      ← 28 × 8
     ▢      ← 28 × 10
   + _____
     ▢

**Remember**
Use patterns of zeros to find the product of multiples of 10.

3 × 4 = 12

3 × 40 = 120        30 × 40 = 1,200

3 × 400 = 1,200     300 × 40 = 12,000

So, on average, a Siberian tiger may eat _____ pounds of food in 28 days.

**Example**

A Siberian tiger was observed sleeping 1,287 minutes during the course of one day. If he slept for that long every day, how many minutes would he sleep in one year? Assume there are 365 days in one year.

**STEP 1** Estimate: 1,287 × 365

Think: 1,000 × 400 = _____

**STEP 2** Multiply by the ones.

 1,287
× 365
_____

1,287 × 5 ones = _____ ones

**STEP 3** Multiply by the tens.

 1,287
× 365
_____

1,287 × 6 tens = _____ tens, or _____ ones

**STEP 4** Multiply by the hundreds.

 1,287
× 365
_____

1,287 × 3 hundreds = _____ hundreds, or _____ ones

**STEP 5** Add the partial products.

 1,287
× 365
_____
  ← 1,287 × 5
  ← 1,287 × 60
+  ← 1,287 × 300
_____

So, the tiger would sleep _____ minutes in one year.

 **Math Talk**

**MATHEMATICAL PRACTICES ⑥**

Are there different numbers you could have used in Step 1 to find an estimate that is closer to the actual answer? **Explain.**

Name _____

**Complete to find the product.**

**1.**

|   |   | 6 | 4 |
|---|---|---|---|
| × |   | 4 | 3 |

← 64 × _____
← 64 × _____
+

**2.**

|   |   | 5 | 7 | 1 |
|---|---|---|---|---|
| × |   |   | 3 | 8 |

← 571 × _____
← 571 × _____
+

**Estimate. Then find the product.**

**3.** Estimate: _____

$$\begin{array}{r} 24 \\ \times\ 15 \\ \hline \end{array}$$

**✓ 4.** Estimate: _____

$$\begin{array}{r} 37 \\ \times\ 63 \\ \hline \end{array}$$

**✓ 5.** Estimate: _____

$$\begin{array}{r} 384 \\ \times\ 45 \\ \hline \end{array}$$

## On Your Own

**Estimate. Then find the product.**

**6.** Estimate: _____

$$\begin{array}{r} 28 \\ \times\ 22 \\ \hline \end{array}$$

**7.** Estimate: _____

$$\begin{array}{r} 93 \\ \times\ 76 \\ \hline \end{array}$$

**8.** Estimate: _____

$$\begin{array}{r} 5,271 \\ \times\ 129 \\ \hline \end{array}$$

**Practice: Copy and Solve** **Estimate. Then find the product.**

**9.** $54 \times 31$     **10.** $42 \times 26$     **11.** $38 \times 64$     **12.** $63 \times 16$

**13.** $204 \times 41$     **14.** $534 \times 25$     **15.** $722 \times 39$     **16.** $957 \times 243$

**17.** **GO DEEPER** One case of books weighs 35 pounds. One case of magazines weighs 23 pounds. A book store wants to ship 72 cases of books and 94 cases of magazines to another store. What is the total weight of the shipment?

## Problem Solving • Applications  (Real World)

**Use the table for 18–20.**

**18.** How much sleep does a jaguar get in 1 year?

_____

**19.** [THINK SMARTER]  In 1 year, how many more hours of sleep does a giant armadillo get than a platypus?

_____

**20.** [MATHEMATICAL PRACTICE 1] **Make Sense of Problems**
Owl monkeys sleep during the day, waking about 15 minutes after sundown to find food. At midnight, they rest for an hour or two, then continue to feed until sunrise. They live about 27 years. How many hours of sleep does an owl monkey that lives 27 years get in its lifetime?

_____

**21.** [GO DEEPER]  Tickets to a museum cost $17 each. For a field trip, the museum offers a $4 discount on each ticket. How much will tickets for 32 students cost?

_____

| Animal Sleep Amounts | |
|---|---|
| **Animal** | **Amount (usual hours per week)** |
| Jaguar | 77 |
| Giant Armadillo | 127 |
| Owl Monkey | 119 |
| Platypus | 98 |
| Three-Toed Sloth | 101 |

[WRITE] *Math* • **Show Your Work**

**22.** [THINK SMARTER]  Rachel earns $21 per day. For 22a–22d, select True or False for each statement.

**22a.** Rachel earns $421 for 20 days of work.

　○ True　　　　　○ False

**22b.** Rachel earns $315 for 15 days of work.

　○ True　　　　　○ False

**22c.** Rachel earns $273 for 13 days of work.

　○ True　　　　　○ False

**22d.** Rachel earns $250 for 13 days of work.

　○ True　　　　　○ False

# Multiply by Multi-Digit Numbers

**Common Core** **COMMON CORE STANDARD—5.NBT.B.5**
*Perform operations with multi-digit whole numbers and with decimals to hundredths.*

## Estimate. Then find the product.

**1.** Estimate: _____4,000_____

```
      82
  ×   49
     738
  + 3280
   4,018
```

**2.** Estimate: _____

```
      92
  ×   68
```

**3.** Estimate: _____

```
   1,537
  ×   242
```

**4.** $23 \times 67$

**5.** $309 \times 29$

**6.** $612 \times 87$

## Problem Solving Real World

**7.** A company shipped 48 boxes of canned dog food. Each box contains 24 cans. How many cans of dog food did the company ship in all?

**8.** There were 135 cars in a rally. Each driver paid a $25 fee to participate in the rally. How much money did the drivers pay in all?

**9.** **WRITE** ▸*Math* Write a problem multiplying a 3-digit number by a 2-digit number. Show all the steps to solve it by using place value and regrouping and by using partial products.

## Lesson Check (5.NBT.B.5)

**1.** A chessboard has 64 squares. At a chess tournament 84 chessboards were used. How many squares are there on 84 chessboards?

**2.** Last month, a manufacturing company shipped 452 boxes of ball bearings. Each box contains 48 ball bearings. How many ball bearings did the company ship last month?

## Spiral Review (5.NBT.A.1, 5.NBT.A.2, 5.NBT.B.5, 5.NBT.B.6)

**3.** What is the standard form of the number three million, sixty thousand, five hundred twenty?

**4.** What number completes the following equation?

$$8 \times (40 + 7) = (8 \times \phantom{xx}) + (8 \times 7)$$

**5.** The population of Clarksville is about 6,000 people. What is the population written as a whole number multiplied by a power of ten?

**6.** A sporting goods store ordered 144 cans of tennis balls. Each can contains 3 balls. How many tennis balls did the store order?

**FOR MORE PRACTICE
GO TO THE
Personal Math Trainer**

Name _____

# Relate Multiplication to Division

**Essential Question** How is multiplication used to solve a division problem?

**Common Core** **Number and Operations in Base Ten—5.NBT.B.6**
**MATHEMATICAL PRACTICES**
**MP2, MP3, MP6, MP7**

You can use the relationship between multiplication and division to solve a division problem. Using the same numbers, multiplication and division are opposite, or **inverse operations.**

$$3 \quad \times \quad 8 \quad = \quad 24 \qquad\qquad 24 \quad \div \quad 3 \quad = \quad 8$$

factor    factor    product       dividend    divisor    quotient

## 🔑 Unlock the Problem Real World

Joel and 5 friends collected 126 marbles. They shared the marbles equally. How many marbles will each person get?

- Underline the dividend.
- What is the divisor? _____

### 🔑 One Way Make an array.

- Outline a rectangular array on the grid to model 126 squares arranged in 6 rows of the same length. Shade each row a different color.

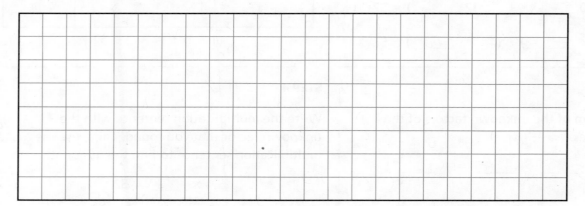

- How many squares are shaded in each row? _____

- Use the array to complete the multiplication sentence. Then, use the multiplication sentence to complete the division sentence.

  $6 \times$ _____ $= 126$         $126 \div 6 =$ _____

So, each of the 6 friends will get _____ marbles.

## 🔒 Another Way Use the Distributive Property.

**Divide.** 52 ÷ 4

You can use the Distributive Property and an area model to solve division problems. Remember that the Distributive Property states that multiplying a sum by a number is the same as multiplying each addend in the sum by the number and then adding the products.

### STEP 1

Write a related multiplication sentence for the division problem.

**Think:** Use the divisor as a factor and the dividend as the product. The quotient will be the unknown factor.

$$52 \div 4 = \blacksquare$$

$$4 \times \blacksquare = 52$$

?

4 | 52

$$4 \times ? = 52$$

### STEP 2

Use the Distributive Property to break apart the large area into smaller areas for partial products that you know.

( 40 + 12 ) = 52

(4 × _____) + (4 × _____) = 52

? ?

4 | 40 | 12

$$(4 \times ?) + (4 \times ?) = 52$$

### STEP 3

Find the sum of the unknown factors of the smaller areas.

_____ + _____ = _____

### STEP 4

Write the multiplication sentence with the unknown factor that you found. Then, use the multiplication sentence to find the quotient.

$$4 \times \underline{\hspace{1cm}} = 52$$

$$52 \div 4 = \underline{\hspace{1cm}}$$

- **MATHEMATICAL PRACTICE 6** **Explain** how you can use the Distributive Property to find the quotient of 96 ÷ 8.

_____

_____

_____

Name _____

## Share and Show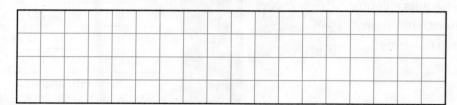

1. Brad has 72 toy cars that he puts into 4 equal groups. How many cars does Brad have in each group? Use the array to show your answer.

$4 \times$ _____ $= 72$   $72 \div 4 =$ _____

|  |  |  |  |  |  |  |  |  |  |  |  |
|--|--|--|--|--|--|--|--|--|--|--|--|
|  |  |  |  |  |  |  |  |  |  |  |  |
|  |  |  |  |  |  |  |  |  |  |  |  |
|  |  |  |  |  |  |  |  |  |  |  |  |

**Use multiplication and the Distributive Property to find the quotient.**

2. $108 \div 6 =$ _____

☑ 3. $84 \div 6 =$ _____

☑ 4. $184 \div 8 =$ _____

**Math Talk**

MATHEMATICAL PRACTICES ⑦

Look for Structure How does using multiplication help you solve a division problem?

## On Your Own

**Use multiplication and the Distributive Property to find the quotient.**

5. $60 \div 4 =$ _____

6. $144 \div 6 =$ _____

7. $252 \div 9 =$ _____

**THINK SMARTER** Find each quotient. Then compare. Write <, >, or =.

8. $51 \div 3$ ◯ $68 \div 4$

9. $252 \div 6$ ◯ $135 \div 3$

10. $110 \div 5$ ◯ $133 \div 7$

## Problem Solving • Applications

**Use the table to solve 11–12.**

11. **THINK SMARTER** Mr. Henderson has 2 bouncy-ball vending machines. He buys one bag of the 27-millimeter balls and one bag of the 40-millimeter balls. He puts an equal number of each in the 2 machines. How many bouncy balls does he put in each machine?

| Bouncy Balls | |
|---|---|
| **Size** | **Number in Bag** |
| 27 mm | 180 |
| 40 mm | 80 |
| 45 mm | 180 |
| mm = millimeters | |

_____

12. **GO DEEPER** Lindsey buys a bag of each size of bouncy ball. She wants to put the same number of each size of bouncy ball into 5 party-favor bags. How many of each size of bouncy ball will she put in a bag?

_____

13. **MATHEMATICAL PRACTICE ③ Verify the Reasoning of Others** Sandra writes $(4 \times 30) + (4 \times 2)$ and says the quotient for $128 \div 4$ is 8. Is she correct? Explain.

**WRITE** ▸ *Math* • **Show Your Work**

_____

_____

_____

_____

**Personal Math Trainer**

14. **THINK SMARTER +** Joe collected 45 seashells. Joe wants to share his seashells with 5 of his friends equally. How many seashells will each friend get? Use the array to show your answer.

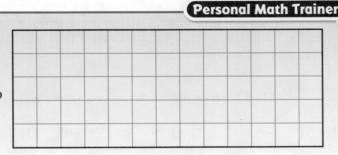

Use the multiplication sentence to complete the division sentence.

$5 \times \boxed{\phantom{0}} = 45$ $\qquad$ $45 \div 5 = \boxed{\phantom{0}}$

Name _____

## Relate Multiplication to Division

Common Core    **Common Core Standard—5.NBT.B.6**
*Perform operations with multi-digit whole numbers and with decimals to hundredths.*

**Use multiplication and the Distributive Property to find the quotient.**

**1.** $70 \div 5 =$ _____14_____

(5 × 10) + (5 × 4) = 70

5 × 14 = 70

**2.** $96 \div 6 =$ _____

_____

_____

**3.** $85 \div 5 =$ _____

_____

_____

**4.** $171 \div 9 =$ _____

_____

_____

**5.** $102 \div 6 =$ _____

_____

_____

**6.** $210 \div 5 =$ _____

_____

_____

## Problem Solving · Real World

**7.** Ken is making gift bags for a party. He has 64 colored pens and wants to put the same number in each bag. How many bags will Ken make if he puts 4 pens in each bag?

_____

**8.** Maritza is buying wheels for her skateboard shop. She ordered a total of 92 wheels. If wheels come in packages of 4, how many packages will she receive?

_____

**9.** **WRITE** ▸*Math*  For the problem $135 \div 5$, draw two different ways to break apart the array. Use the Distributive Property to write products for each different way.

_____

_____

_____

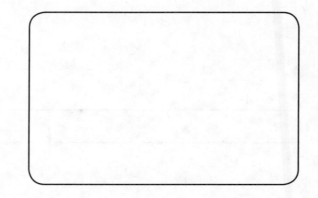

## Lesson Check (5.NBT.B.6)

**1.** Write an expression using the Distributive Property that can be used to find the quotient of 36 ÷ 3.

**2.** Write an expression using the Distributive Property that can be used to find the quotient of 126 ÷ 7.

## Spiral Review (4.OA.A.3, 5.NBT.A.1, 5.NBT.A.2)

**3.** Allison separates 23 stickers into 4 equal piles. How many stickers does she have left over?

**4.** A website had 2,135,789 hits. What is the value of the digit 3?

**5.** The area of Arizona is 114,006 square miles. What is the expanded form of this number?

**6.** What is the value of the fourth power of ten?

© Houghton Mifflin Harcourt Publishing Company

FOR MORE PRACTICE
GO TO THE
**Personal Math Trainer**

# Problem Solving • Multiplication and Division

**Essential Question** How can you use the strategy *solve a simpler problem* to help you solve a division problem?

 **Common Core** Number and Operations in Base Ten—5.NBT.B.6

MATHEMATICAL PRACTICES
MP1, MP2, MP3

 **Unlock the Problem** Real World

Mark works at an animal shelter. To feed 9 dogs, Mark empties eight 18-ounce cans of dog food into a large bowl. If he divides the food equally among the dogs, how many ounces of food will each dog get?

Use the graphic organizer below to help you solve the problem.

## Read the Problem

**What do I need to find?**

I need to find _____

_____.

**What information do I need to use?**

I need to use the number of _____ , the

number of _____ in each can, and the

number of dogs that need to be fed.

**How will I use the information?**

I can _____ to find the total number of

ounces. Then I can solve a simpler problem to

_____ that total by 9.

## Solve the Problem

- First, multiply to find the total number of ounces of dog food.

  $8 \times 18 =$ _____

- To find the number of ounces each dog gets, I'll need to divide.

  $144 \div$ _____ $=$ ■

- To find the quotient, I break 144 into two simpler numbers that are easier to divide.

  $144 \div 9$ $=$ ■

  $(90 +$ ____$) \div 9$ $=$ ■

  $($____$\div 9) + ($____$\div 9)$ $=$ ■

  ____ $+$ 6 $=$ ____

So, each dog gets _____ ounces of food.

# ❶ Try Another Problem

Michelle is building shelves for her room. She has a plank 137 inches long that she wants to cut into 7 shelves of equal length. The plank has jagged ends, so she will start by cutting 2 inches off each end. How long will each shelf be?

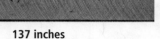

**137 inches**

| Read the Problem | Solve the Problem |
|---|---|
| **What do I need to find?** | |
| **What information do I need to use?** | |
| **How will I use the information?** | |

So, each shelf will be _____ inches long.

**Math Talk**

MATHEMATICAL PRACTICES ❶

**Analyze** Explain how the strategy you used helped you solve the problem.

Name _____

# Unlock the Problem

✓ Underline what you need to find.
✓ Circle the numbers you need to use.

## Share and Show

1. To make concrete mix, Monica pours 34 pounds of cement, 68 pounds of sand, 14 pounds of small pebbles, and 19 pounds of large pebbles into a large wheelbarrow. If she pours the mixture into 9 equal-size bags, how much will each bag weigh?

   **First,** find the total weight of the mixture.

   _____

   **Then,** divide the total by the number of bags. Break the total into two simpler numbers to make the division easier, if necessary.

   WRITE ▸*Math* • Show Your Work • • • •

   _____

   **Finally,** find the quotient and solve the problem.

   So, each bag will weigh _____ pounds.

2. **What if** Monica pours the mixture into 5 equal-size bags? How much will each bag weigh?

   _____

✓ 3. Taylor is building doghouses to sell. Each doghouse requires 3 full sheets of plywood which Taylor cuts into new shapes. The plywood is shipped in bundles of 14 full sheets. How many doghouses can Taylor make from 12 bundles of plywood?

   _____

✓ 4. Eileen is planting a garden. She has seeds for 60 tomato plants, 55 sweet corn plants, and 21 cucumber plants. She plants them in 8 rows, with the same number of plants in each row. How many seeds are planted in each row?

   _____

## On Your Own

5. **GO DEEPER** Starting on day 1 with 1 jumping jack, Keila doubles the number of jumping jacks she does every day. How many jumping jacks will Keila do on day 10?

_____

6. **MATHEMATICAL PRACTICE 2** **Represent a Problem** Starting in the blue square, in how many different ways can you draw a line that passes through every square without picking up your pencil or crossing a line you've already drawn? Show the ways.

_____

7. On April 11, Millie bought a lawn mower with a 50-day guarantee. If the guarantee begins on the date of purchase, what is the first day on which the mower will no longer be guaranteed?

_____

8. **THINK SMARTER** The teacher of a jewelry-making class had a supply of 236 beads. Her students used 29 beads to make earrings and 63 beads to make bracelets. They will use the remaining beads to make necklaces with 6 beads on each necklace. How many necklaces will the students make?

_____

9. **THINK SMARTER** Susan is making 8 casseroles. She uses 9 cans of beans. Each can is 16-ounces. If she divides the beans equally among 8 casseroles, how many ounces of beans will be in each casserole? Show your work.

_____

_____

_____

## Problem Solving • Multiplication and Division

**COMMON CORE STANDARD—5.NBT.B.6**
*Perform operations with multi-digit whole numbers and with decimals to hundredths.*

**Solve the problems below. Show your work.**

1. Dani is making punch for a family picnic. She adds 16 fluid ounces of orange juice, 16 fluid ounces of lemon juice, and 8 fluid ounces of lime juice to 64 fluid ounces of water. How many 8-ounce glasses of punch can she fill?

   16 + 16 + 8 + 64 = 104 fluid ounces

   $$104 \div 8 = (40 + 64) \div 8$$
   $$= (40 \div 8) + (64 \div 8)$$
   $$= 5 + 8, \text{ or } 13$$

   _____ **13 glasses**

2. Ryan has nine 14-ounce bags of popcorn to repackage and sell at the school fair. A small bag holds 3 ounces. How many small bags can he make?

   _____

3. Bianca is making scarves to sell. She has 33 pieces of blue fabric, 37 pieces of green fabric, and 41 pieces of red fabric. Suppose Bianca uses 3 pieces of fabric to make 1 scarf. How many scarves can she make?

   _____

4. Jasmine has 8 packs of candle wax to make scented candles. Each pack contains 14 ounces of wax. Jasmine uses 7 ounces of wax to make one candle. How many candles can she make?

   _____

5. **WRITE** *Math* Rewrite Problem 4 on page 57 with different numbers. Solve the new problem and show your work.

   _____

   _____

   _____

## Lesson Check (5.NBT.B.6)

**1.** Joyce is helping her aunt create craft kits. Her aunt has 138 pipe cleaners, and each kit will include 6 pipe cleaners. What is the total number of craft kits they can make?

_____

**2.** Stefan plants seeds for 30 carrot plants and 45 beet plants in 5 rows, with the same number of seeds in each row. How many seeds are planted in each row?

_____

## Spiral Review (Reviews 4.NBT.A.3, 5.NBT.B.5, 5.NBT.B.6)

**3.** Georgia wants to evenly divide 84 trading cards among 6 friends. How many cards will each friend get?

_____

**4.** Maria has 144 marbles. Emanuel has 4 times the number of marbles Maria has. How many marbles does Emanuel have?

_____

**5.** The Conservation Society bought and planted 45 cherry trees. Each tree cost $367. What was the total cost of planting the trees?

_____

**6.** A sports arena covers 710,430 square feet of ground. A newspaper reported that the arena covers about 700,000 square feet of ground. To what place value was the number rounded?

_____

© Houghton Mifflin Harcourt Publishing Company

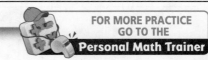

FOR MORE PRACTICE
GO TO THE
Personal Math Trainer

Name _____

# Numerical Expressions

**Essential Question** How can you use a numerical expression to describe a situation?

 **Operations and Algebraic Thinking—5.OA.A.1, 5.OA.A.2**
**MATHEMATICAL PRACTICES**
MP3, MP4, MP6

 **Unlock the Problem** Real World

A **numerical expression** is a mathematical phrase that has numbers and operation signs but does not have an equal sign.

Tyler caught 15 small bass, and his dad caught 12 small bass in the Memorial Bass Tourney in Tidioute, PA. Write a numerical expression to represent how many fish they caught in all.

**Choose which operation to use.**

You need to join groups of different sizes, so use addition.

| 15 small bass | plus | 12 small bass |
| :---: | :---: | :---: |
| ↓ | ↓ | ↓ |
| 15 | + | 12 |

So, 15 + 12 represents how many fish they caught in all.

# Example 1 Write an expression to match the words.

### A Addition

Emma has 11 fish in her aquarium. She buys 4 more fish.

| fish | plus | more fish |
| :---: | :---: | :---: |
| ↓ | ↓ | ↓ |
| 11 | + | 4 |

### B Subtraction

Lucia has 128 stamps. She uses 38 stamps on party invitations.

| stamps | minus | stamps used |
| :---: | :---: | :---: |
| ↓ | ↓ | ↓ |
| 128 | − | _____ |

### C Multiplication

Karla buys 5 books. Each book costs $3.

| books | multiplied by | cost per book |
| :---: | :---: | :---: |
| ↓ | ↓ | ↓ |
| _____ | × | _____ |

### D Division

Four players share 52 cards equally.

| cards | divided by | players |
| :---: | :---: | :---: |
| ↓ | ↓ | ↓ |
| _____ | ÷ | _____ |

 Math Talk

**MATHEMATICAL PRACTICES** ④

What does the expression model in each example?

**Expressions with Parentheses** The meaning of the words in a problem will tell you where to place the parentheses in an expression.

 **Example 2** Which expression matches the meaning of the words?

Doug went fishing for 3 days. Each day he put $15 in his pocket. At the end of each day, he had $5 left. How much money did Doug spend by the end of the trip?

> • Underline the events for each day.
> • Circle the number of days these events happened.

**Think:** Each day he took $15 and had $5 left. He did this for 3 days.

($15 − $5)    ← **Think:** What expression can you write to show how much money Doug spends in one day?

3 × ($15 − $5)    ← **Think:** What expression can you write to show how much money Doug spends in three days?

 **Math Talk**

MATHEMATICAL PRACTICES ③

Explain how the expression of what Doug spent in three days compares to the expression of what he spent in one day?

 **Example 3** Which problem matches the expression $20 − ($12 + $3)?

Kim has $20 to spend for her fishing trip. She spends $12 on a fishing pole. Then she finds $3. How much money does Kim have now?

List the events in order.

First: Kim has $20.

Next: _____.

Then: _____.

Do these words match the expression? _____

Kim has $20 to spend for her fishing trip. She spends $12 on a fishing pole and $3 on bait. How much money does Kim have now?

List the events in order.

First: Kim has $20.

Next: _____.

Then: _____.

Do these words match the expression? _____

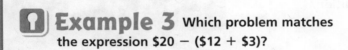 **Share and Show** MATH BOARD

**Circle the expression that matches the words.**

1. Teri had 18 worms. She gave 4 worms to Susie and 3 worms to Jamie.

    (18 − 4) + 3        18 − (4 + 3)

2. Rick had $8. He then worked 4 hours for $5 each hour.

    $8 + (4 × $5)        ($8 + 4) × $5

Name _____

**Write an expression to match the words.**

**3.** Greg drives 26 miles on Monday and 90 miles on Tuesday.

_____

**✓ 4.** Lynda has 27 fewer fish than Jack. Jack has 80 fish.

_____

**Write words to match the expression.**

**5.** $34 - 17$

_____

_____

**✓ 6.** $6 \times (12 - 4)$

_____

_____

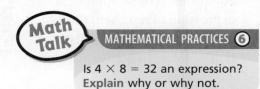
Math Talk    MATHEMATICAL PRACTICES ⑥

Is $4 \times 8 = 32$ an expression? **Explain** why or why not.

## On Your Own

**Write an expression to match the words.**

**7.** José shared 12 party favors equally among 6 friends.

_____

**8.** Braden has 14 baseball cards. He finds 5 more baseball cards.

_____

**9.** Isabelle bought 12 bottles of water at $2 each.

_____

**10.** Monique had $20. She spent $5 on lunch and $10 at the bookstore.

_____

**Write words to match the expression.**

**11.** $36 \div 9$

_____

_____

**12.** $35 - (16 + 11)$

_____

_____

**Draw a line to match the expression with the words.**

**13.** Fred catches 25 fish. Then he releases 10 fish and catches 8 more.  •

Nick has 25 pens. He gives 10 pens to one friend and 8 pens to another friend.  •

Jan catches 15 fish and lets 6 fish go.  •

Libby catches 15 fish and lets 6 fish go for three days in a row.  •

•  $3 \times (15 - 6)$

•  $15 - 6$

•  $25 - (10 + 8)$

•  $(25 - 10) + 8$

**Chapter 1 • Lesson 10   63**

## Problem Solving • Applications

**Use the rule and the table for 14–15.**

14. **MATHEMATICAL PRACTICE 4** **Write an Expression**
to represent the total number of lemon
tetras that could be in a 20-gallon aquarium.

_____

15. **THINK SMARTER** There are tiger
barbs in a 15-gallon aquarium
and giant danios in a 30-gallon
aquarium. Write a numerical
expression to represent the
greatest total number of fish
that could be in both aquariums.

_____

| Aquarium Fish | |
| --- | --- |
| **Type of Fish** | **Length (in inches)** |
| Lemon Tetra | 2 |
| Strawberry Tetra | 3 |
| Giant Danio | 5 |
| Tiger Barb | 3 |
| Swordtail | 5 |

▲ The rule for the number of fish in
an aquarium is to allow 1 gallon
of water for each inch of length.

16. **GO DEEPER** Write a word problem for an
expression that is three times as great as
(15 + 7). Then write the expression.

_____

_____

_____

_____

17. **THINK SMARTER** Daniel bought 30 tokens when
he arrived at the festival. He won 8 more
tokens for getting the highest score at the
basketball contest, but lost 6 tokens at the ring
toss game. Write an expression to find the
number of tokens Daniel has left.

_____

| WRITE ▸ Math • **Show Your Work**

## Numerical Expressions

Common Core

COMMON CORE STANDARD—5.OA.A.1,
5.OA.A.2 *Write and interpret numerical
expressions.*

**Write an expression to match the words.**

**1.** Ethan collected 16 seashells. He lost 4 of them while walking home.

_____ 16 − 4 _____

**2.** Yasmine bought 4 bracelets. Each bracelet cost $3.

_____

**3.** Amani did 10 jumping jacks. Then she did 7 more.

_____

**4.** Darryl has a board that is 8 feet long. He cuts it into pieces that are each 2 feet long.

_____

**Write words to match the expression.**

**5.** $3 + (4 \times 12)$

_____

_____

_____

**6.** $36 \div 4$

_____

_____

_____

**7.** $24 − (6 + 3)$

_____

_____

_____

**Problem Solving** Real World

**8.** Kylie has 14 polished stones. Her friend gives her 6 more stones. Write an expression to match the words.

_____

**9.** Rashad had 25 stamps. He shared them equally among himself and 4 friends. Then Rashad found 2 more stamps in his pocket. Write an expression to match the words.

_____

**10.** **WRITE** ▸ *Math* Write a numerical expression. Then write words to match the expression.

_____

_____

## Lesson Check (5.OA.A.1)

**1.** Jenna bought 3 packs of bottled water, with 8 bottles in each pack. Then she gave 6 bottles away. Write an expression to match the words.

_____

**2.** Stephen had 24 miniature cars. He gave 4 cars to his brother. Then he passed the rest of the cars out equally among 4 of his friends. Which operation would you use to represent the first part of this situation?

_____

## Spiral Review (5.NBT.A.2, 5.NBT.B.5, 5.NBT.B.6)

**3.** To find $36 + 29 + 14$, Joshua rewrote the expression as $36 + 14 + 29$. What property did Joshua use to rewrite the expression?

_____
_____

**4.** There are 6 baskets on the table. Each basket has 144 crayons in it. How many crayons are there?

_____
_____

**5.** Mr. Anderson wrote $(7 \times 9) \times 10^3$ on the board. What is the value of that expression?

_____

**6.** Barbara mixes 54 ounces of granola and 36 ounces of raisins. She divides the mixture into 6-ounce servings. How many servings does she make?

_____

FOR MORE PRACTICE
GO TO THE
**Personal Math Trainer**

Name _____

# Evaluate Numerical Expressions

**Essential Question** In what order must operations be evaluated to find the solution to a problem?

**Common Core** Operations and Algebraic Thinking—5.OA.A.1
**MATHEMATICAL PRACTICES**
**MP2, MP3, MP4**

**CONNECT** Remember that a numerical expression is a mathematical phrase that uses only numbers and operation symbols.

$(5 - 2) \times 7$ $\qquad$ $72 \div 9 + 16$ $\qquad$ $(24 - 15) + 32$

To **evaluate**, or find the value of, a numerical expression with more than one type of operation, you must follow rules called the **order of operations.** The order of operations tells you in what order you should evaluate an expression.

> **Order of Operations**
> 1. Perform operations in parentheses.
> 2. Multiply and divide from left to right.
> 3. Add and subtract from left to right.

## Unlock the Problem

A bread recipe calls for 4 cups of wheat flour and 2 cups of rye flour. To triple the recipe, how many cups of flour are needed in all?

 Evaluate **3 × 4 + 3 × 2** to find the total number of cups.

**A** Gabriela did not follow the order of operations correctly.

| | Gabriela |
|---|---|
| ○ 3 × 4 + 3 × 2 | First, I added. |
| ○ 3 × 7 × 2 | Then, I multiplied. |
| 42 | |

**Explain** why Gabriela's answer is not correct.

_____

**B** Follow the order of operations by multiplying first and then adding.

| | Name_____ |
|---|---|
| ○ 3 × 4 + 3 × 2 | |
| ○ | |
| | |

So, _____ cups of flour are needed.

**Evaluate Expressions with Parentheses** To evaluate an expression with parentheses, follow the order of operations. Perform the operations in parentheses first. Multiply from left to right. Then add and subtract from left to right.

## 🔒 Example

Each batch of granola Lena makes uses 3 cups of oats, 1 cup of raisins, and 2 cups of nuts. Lena wants to make 5 batches of granola. How many cups of oats, raisins, and nuts will she need in all?

Write the expression.                                     $5 \times (3 + 1 + 2)$

First, perform the operations in parentheses.        $5 \times ($_____$)$

Then multiply.                                                    _____

So, Lena will use _____ cups of oats, raisins, and nuts in all.

---

• **MATHEMATICAL PRACTICE ②** **Reason Quantitatively** What if Lena makes 4 batches? Will this change the numerical expression? Explain.

_____

---

**Try This!** **Rewrite the expression with parentheses to equal the given value.**

**Ⓐ** $6 + 12 \times 8 - 3$; value: 141

• Evaluate the expression without the parentheses. _____

• Try placing the parentheses in the expression so the value is 141.

  **Think:** Will the placement of the parentheses increase or decrease the value of the expression?

• Use order of operations to check your work.

  $6 + 12 \times 8 - 3$

---

**Ⓑ** $5 + 28 \div 7 - 4$; value: 11

• Evaluate the expression without the parentheses. _____

• Try placing the parentheses in the expression so that the value is 11.

  **Think:** Will the placement of the parentheses increase or decrease the value of the expression?

• Use order of operations to check your work.

  $5 + 28 \div 7 - 4$

Name _____

**Evaluate the numerical expression.**

**1.** $10 + 36 \div 9$

Think: I need to divide first.

_____

**⊘ 2.** $10 + (25 - 10) \div 5$

_____

**⊘ 3.** $9 - (3 \times 2) + 8$

_____

## On Your Own

**Math Talk**

**MATHEMATICAL PRACTICES ③**

Raina evaluated the expression $5 \times 2 + 2$ by adding first and then multiplying. Will her answer be correct? **Apply** the order of operations.

**Evaluate the numerical expression.**

**4.** $(4 + 49) - 4 \times 10$

_____

**5.** $5 + 17 - 100 \div 5$

_____

**6.** $36 - (8 + 5)$

_____

**7.** $125 - (68 + 7)$

_____

**Rewrite the expression with parentheses to equal the given value.**

**8.** $100 - 30 \div 5$
value: 14

_____

**9.** $12 + 17 - 3 \times 2$
value: 23

_____

**10.** $9 + 5 \div 5 + 2$
value: 2

_____

**11.** **THINK SMARTER** Each pitcher of power smoothie that Ginger makes has 2 scoops of pineapple, 3 scoops of strawberries, 1 scoop of spinach, and 1 scoop of kale. If Ginger makes 7 pitchers of power smoothies, how many scoops will she use in all? Write and evaluate a numerical expression containing parentheses.

_____

_____

**12.** **MATHEMATICAL PRACTICE ②** **Reason Abstractly** The value of $100 - 30 \div 5$ with parentheses can have a value of 14 or 94. Explain.

_____

_____

_____

## Unlock the Problem  Real World

**13.** *GO DEEPER*  A movie theater has 4 groups of seats. The largest group of seats, in the middle, has 20 rows, with 20 seats in each row. There are 2 smaller groups of seats on the sides, each with 20 rows and 6 seats in each row. A group of seats in the back has 5 rows, with 30 seats in each row. How many seats are in the movie theater?

| back |
| side | middle | side |

**a.** What do you need to know? _____

_____

**b.** What operation can you use to find the number of seats in the back

group of seats? Write the expression. _____

**c.** What operation can you use to find the number of seats in both groups of side seats? Write the expression.

_____

**d.** What operation can you use to find the number of seats in the middle group? Write the expression.

_____

**e.** Write an expression to represent the total number of seats in the theater.

_____

_____

**f.** How many seats are in the theater? Show the steps you use to solve the problem.

_____

_____

**14.** *THINK SMARTER*

Write and evaluate two equivalent numerical expressions that show the Distributive Property of Multiplication.

Math on the Spot

_____

_____

**15.** *THINK SMARTER*  Rosalie evaluates the numerical expression $4 + 5 \times 2 - 1$.

Rosalie's first step should be to

| add |
| subtract |
| multiply |

.

**Practice and Homework**
**Lesson 1.11**

## Evaluate Numerical Expressions

**COMMON CORE STANDARD—5.OA.A.1**
*Write and interpret numerical expressions.*

**Evaluate the numerical expression.**

**1.** $24 \times 5 - 41$
$120 - 41$

$\underline{\qquad 79 \qquad}$

**2.** $(32 - 20) \div 4$

**3.** $16 \div (2 + 6)$

**4.** $27 + 5 \times 6$

**Rewrite the expression with parentheses to equal the given value.**

**5.** $3 \times 4 - 1 + 2$

value: 11

**6.** $2 \times 6 \div 2 + 1$

value: 4

**7.** $5 + 3 \times 2 - 6$

value: 10

## Problem Solving · Real World

**8.** Sandy has several pitchers to hold lemonade for the school bake sale. Two pitchers can hold 64 ounces each, and four pitchers can hold 48 ounces each. How many total ounces can Sandy's pitchers hold?

_____

**9.** At the bake sale, Jonah sold 4 cakes for $8 each and 36 muffins for $2 each. What was the total amount, in dollars, that Jonah received from these sales?

_____

**10.** **WRITE** ▸*Math* Give two examples that show how using parentheses can change the order in which operations are performed in an expression.

_____

_____

## Lesson Check (5.OA.A.1)

**1.** What is the value of the expression $4 \times (4 - 2) + 6$?

**2.** Lannie ordered 12 copies of the same book for his book club members. The books cost $19 each, and the order has a $15 shipping charge. What is the total cost of Lannie's order?

_____

_____

## Spiral Review (5.NBT.A.1, 5.NBT.A.2, 5.NBT.B.5, 5.NBT.B.6)

**3.** A small company packs 12 jars of jelly into each of 110 boxes to bring to the farmers' market. How many jars of jelly does the company pack in all?

**4.** June has 42 sports books, 85 mystery books, and 69 nature books. She arranges her books equally on 7 shelves. How many books are on each shelf?

_____

_____

**5.** Last year, a widget factory produced one million, twelve thousand, sixty widgets. What is this number written in standard form?

**6.** A company has 3 divisions. Last year, each division earned a profit of $\$5 \times 10^5$. What was the total profit the company earned last year?

_____

_____

FOR MORE PRACTICE
GO TO THE
**Personal Math Trainer**

# Grouping Symbols

**Essential Question** In what order must operations be evaluated to find a solution when there are parentheses within parentheses?

 **Operations and Algebraic Thinking—5.OA.A.1**

**MATHEMATICAL PRACTICES**
**MP2, MP4**

 **Unlock the Problem**

Mary's weekly allowance is $8 and David's weekly allowance is $5. Every week they each spend $2 on lunch. Write a numerical expression to show how many weeks it will take them together to save enough money to buy a video game for $45.

- Underline Mary's weekly allowance and how much she spends.
- Circle David's weekly allowance and how much he spends.

🔑 **Use parentheses and brackets to write an expression.**

You can use parentheses and brackets to group operations that go together. Operations in parentheses and brackets are performed first.

**STEP 1** Write an expression to represent how much Mary and David save each week.

- How much money does Mary save each week?

  **Think:** Each week Mary gets $8 and spends $2.

  ( _____ )

- How much money does David save each week?

  **Think:** Each week David gets $5 and spends $2.

  ( _____ )

- How much money do Mary and David save together each week? _____

**STEP 2** Write an expression to represent how many weeks it will take Mary and David to save enough money for the video game.

- How many weeks will it take Mary and David to save enough for a video game?

  **Think:** I can use brackets to group operations a second time. $45 is divided by the total amount of money saved each week.

  _____ ÷ [ _____ ]

**Math Talk**

**MATHEMATICAL PRACTICES ④**

**Modeling** Explain why brackets are placed around the part of the expression that represents the amount of money Mary and David save each week.

**Evaluate Expressions with Grouping Symbols** When evaluating an expression with different grouping symbols (parentheses, brackets, and braces), perform the operation in the innermost set of grouping symbols first, evaluating the expression from the inside out.

## 🔑 Example

Juan gets $6 for his weekly allowance and spends $4 of it. His sister Tina gets $7 for her weekly allowance and spends $3 of it. Their mother's birthday is in 4 weeks. If they spend the same amount each week, how much money can they save together in that time to buy her a present?

* Write the expression using parentheses and brackets.

    $4 \times [(\$6 - \$4) + (\$7 - \$3)]$

* Perform the operations in the parentheses first.

    $4 \times [\underline{\hspace{1cm}} + \underline{\hspace{1cm}}]$

* Next perform the operations in the brackets.

    $4 \times \underline{\hspace{1cm}}$

* Then multiply.

    $\underline{\hspace{1cm}}$

So, Juan and Tina will be able to save $\underline{\hspace{2cm}}$ for their mother's birthday present.

---

* **MATHEMATICAL PRACTICE 2** **Connect Symbols and Words** What if only Tina saves any money? Will this change the numerical expression? Explain.

$\underline{\hspace{15cm}}$

**Try This!** Follow the order of operations.

**A** $4 \times \{[(5 - 2) \times 3] + [(2 + 4) \times 2]\}$

* Perform the operations in the parentheses.    $4 \times \{[3 \times 3] + [\underline{\hspace{1cm}} \times \underline{\hspace{1cm}}]\}$

* Perform the operations in the brackets.    $4 \times \{9 + \underline{\hspace{1cm}}\}$

* Perform the operations in the braces.    $4 \times \underline{\hspace{1cm}}$

* Multiply.    $\underline{\hspace{1cm}}$

**B** $32 \div \{[(3 \times 2) + 7] - [(6 - 4) + 7]\}$

* Perform the operations in the parentheses.    $32 \div \{[\underline{\hspace{1cm}} + \underline{\hspace{1cm}}] - [\underline{\hspace{1cm}} + \underline{\hspace{1cm}}]\}$

* Perform the operations in the brackets.    $32 \div \{\underline{\hspace{1cm}} - \underline{\hspace{1cm}}\}$

* Perform the operations in the braces.    $32 \div \underline{\hspace{1cm}}$

* Divide.    $\underline{\hspace{1cm}}$

Name _____

**Evaluate the numerical expression.**

**1.** $12 + [(15 - 5) + (9 - 3)]$

   $12 + [10 + _____]$

   $12 + _____$

   _____

✓ **2.** $5 \times [(26 - 4) - (4 + 6)]$

   _____

✓ **3.** $36 \div [(18 - 10) - (8 - 6)]$

   _____

## On Your Own

**Evaluate the numerical expression.**

**4.** $4 + [(16 - 4) + (12 - 9)]$

**5.** $24 - [(10 - 7) + (16 - 9)]$

**6.** $3 \times \{[(12 - 8) \times 2] + [(11 - 9) \times 3]\}$

## Problem Solving • Applications

**7.** **MATHEMATICAL PRACTICE ④** **Use Symbols** Write the expression $2 \times 8 + 20 - 12 \div 6$ with parentheses and brackets two different ways so one value is less than 10 and the other value is greater than 50.

_____

**8.** **GO DEEPER** Wilma works at a bird sanctuary and stores birdseed in plastic containers. She has 3 small containers that hold 8 pounds of birdseed each and 6 large containers that hold 12 pounds of birdseed each. Each container was full until she used 4 pounds of bird seed. She wants to put some of the remaining birdseed into 30 bird feeders that can hold 2 pounds each. How much birdseed does she have left over? Show the expression you used to find your answer.

_____

## 🔑 Unlock the Problem

9. **THINK SMARTER** Dan has a flower shop. Each day he displays 24 roses. He gives away 10 and sells the rest. Each day he displays 36 carnations. He gives away 12 and sells the rest. What expression can you use to find out how many roses and carnations Dan sells in a week?

a. What information are you given? _____

_____

_____

b. What are you being asked to do? _____

_____

_____

c. What expression shows how many roses Dan sells in one day? _____

d. What expression shows how many carnations Dan sells in one day? _____

e. Write an expression to represent the total number

of roses and carnations Dan sells in one day. _____

f. Write the expression that shows how many

roses and carnations Dan sells in a week. _____

**Personal Math Trainer**

10. **THINK SMARTER +** A gift shop had 500 coloring pencils. The shop sold 3 sets of 20 coloring pencils, 6 sets of 12 coloring pencils, and 10 sets of 18 coloring pencils. Write a numerical expression to show how many coloring pencils are left. Evaluate the numerical expression using order of operations. Show your work.

_____

_____

## Grouping Symbols

Common Core
**COMMON CORE STANDARD—5.OA.A.1**
*Write and interpret numerical expressions.*

**Evaluate the numerical expression.**

**1.** $5 \times [(11 - 3) - (13 - 9)]$

$5 \times [8 - (13 - 9)]$

$5 \times [8 - 4]$

$5 \times 4$

_____ 20 _____

**2.** $30 - [(9 \times 2) - (3 \times 4)]$

_____

**3.** $[(25 - 11) + (15 - 9)] \div 5$

_____

**4.** $8 \times \{[(7 + 4) \times 2] - [(11 - 7) \times 4]\}$

_____

**5.** $\{[(8 - 3) \times 2] + [(5 \times 6) - 5]\} \div 5$

_____

## Problem Solving • Real World

**Use the information at the right for 6 and 7.**

**6.** Write an expression to represent the total number of muffins and bagels Joan sells in 5 days.

_____

Joan has a cafe. Each day, she bakes 24 muffins. She gives away 3 and sells the rest. Each day, she also bakes 36 bagels. She gives away 4 and sells the rest.

**7.** Evaluate the expression to find the total number of muffins and bagels Joan sells in 5 days.

_____

**8.** **WRITE** ▸ *Math* Explain how to use grouping symbols to organize information appropriately.

_____

_____

## Lesson Check (5.OA.A.1)

**1.** What is the value of the expression?

$$30 + [(6 \div 3) + (3 + 4)]$$

**2.** Find the value of the following expression.

$$[(17 - 9) \times (3 \times 2)] \div 2$$

## Spiral Review (5.OA.A.2, 5.NBT.A.1, 5.NBT.B.5)

**3.** What is $\frac{1}{10}$ of 200?

**4.** The Park family is staying at a hotel near an amusement park for 3 nights. The hotel costs $129 per night. How much will their 3-night stay in the hotel cost?

**5.** Vidal bought 2 pizzas and cut each into 8 slices. He and his friends ate 10 slices. Write an expression to match the words.

**6.** What is the value of the underlined digit in 783,5<u>4</u>9,201?

FOR MORE PRACTICE
GO TO THE
**Personal Math Trainer**

Name _____

1. Find the property that each equation shows.
   Write the equation in the correct box.

$15 \times (7 \times 9) = (15 \times 7) \times 9$

$23 + 4 + 109 = 4 + 23 + 109$

$13 + (3 + 7) = (13 + 3) + 7$

$87 \times 3 = 3 \times 87$

$1 \times 9 = 9$

$0 + 16 = 16$

| Identity Property of Addition | Commutative Property of Multiplication | Identity Property of Multiplication |
|---|---|---|
| Associative Property of Multiplication | Commutative Property of Addition | Associative Property of Addition |

2. For 2a–2d, select True or False for each statement.

   2a.  170 is $\frac{1}{10}$ of 17  ○ True  ○ False

   2b.  660 is 10 times as much as 600  ○ True  ○ False

   2c.  900 is $\frac{1}{10}$ of 9,000  ○ True  ○ False

   2d.  4,400 is 10 times as much as 440  ○ True  ○ False

**GO DIGITAL**  **Assessment Options**
**Chapter Test**

3. Select other ways to write 700,562. Mark all that apply.

   (A) $(7 \times 100,000) + (5 \times 1,000) + (6 \times 10) + (2 \times 1)$

   (B) seven hundred thousand, five hundred sixty-two

   (C) $700,000 + 500 + 60 + 2$

   (D) 7 hundred thousands + 5 hundreds + 62 tens

4. Carrie has 140 coins. She has 10 times as many coins as she had last month. How many coins did Carrie have last month?

   _____ coins

5. Valerie earns $24 per hour. Which expression can be used to show how much money she earns in 7 hours?

   (A) $(7 + 20) + (7 + 4)$

   (B) $(7 \times 20) + (7 \times 4)$

   (C) $(7 + 20) \times (7 + 4)$

   (D) $(7 \times 20) \times (7 \times 4)$

6. The table shows the equations Ms. Valez discussed in math class today.

   | Equations |
   | --- |
   | $6 \times 10^0 = 6$ |
   | $6 \times 10^1 = 60$ |
   | $6 \times 10^2 = 600$ |
   | $6 \times 10^3 = 6,000$ |

   Explain the pattern of zeros in the product when multiplying by powers of 10.

© Houghton Mifflin Harcourt Publishing Company

**7.** It is 3,452 miles round trip to Craig's aunt's house. If he travels to her house 3 times this year, how many miles did he travel in all?

_____ miles

**8.** Lindsey earns $33 per day at her part-time job. Complete the table to show the total amount Lindsey earns.

| Lindsey's Earnings | |
|---|---|
| Number of Days | Total Amount |
| 3 | |
| 8 | |
| 14 | |

**Personal Math Trainer**

**9.** **THINK SMARTER +** Jackie followed these steps to evaluate the expression $15 - (37 + 8) \div 3$.

$37 + 8 = 45$

$45 - 15 = 30$

$30 \div 3 = 10$

Mark looks at Jackie's work and says she made a mistake. He says she should have divided by 3 before she subtracted.

**Part A**

Which student is correct? Explain how you know.

**Part B**

Evaluate the expression.

**10.** Carmine buys 8 plates for $1 each. He also buys 4 bowls. Each bowl costs twice as much as each plate. The store is having a sale that gives Carmine $3 off the bowls. Which numerical expression shows how much he spent?

(A) $(8 \times 1) + [(4 \times 16) - 3]$

(B) $(8 \times 1) + [4 \times (16 - 3)]$

(C) $(8 \times 1) + [(4 \times 2) - 3]$

(D) $(8 \times 4) + [(4 \times 2) - 3]$

**11.** Evaluate the numerical expression.

$2 + (65 + 7) \times 3 = \boxed{\phantom{000}}$

**12.** An adult elephant eats about 300 pounds of food each day. Write an expression to represent the number of pounds of food a herd of 12 elephants eats in 5 days.

$$\boxed{\phantom{00000000000000000000000000000000}}$$

**13.** Jason is solving a homework problem.

Arianna buys 5 boxes of granola bars. Each box contains 12 granola bars. Arianna eats 4 bars.

Jason writes a numerical expression to represent the situation. His expression, $(12 - 4) \times 5$, has a mistake.

**Part A**

Explain Jason's mistake.

**Part B**

Write an expression to show how many granola bars are left, and then solve it.

**14.** Paula collected 75 stickers. She shares her stickers with 5 of her friends equally. How many stickers will each friend get?

**Part A**

Use the array to show your answer.

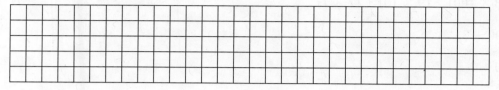

**Part B**

Use the multiplication sentence to complete the division sentence.

$5 \times \boxed{\phantom{00}} = 75$         $75 \div 5 = \boxed{\phantom{00}}$

**15.**  Mario is making dinner for 9 people. Mario buys 6 containers of soup. Each container is 18 ounces. If everyone gets the same amount of soup, how much soup will each person get? How can you solve a simpler problem to help you find the solution?

**16.** Jill wants to find the quotient. Use multiplication and the Distributive Property to help Jill find the quotient.

$144 \div 8 = \boxed{\phantom{00}}$

Multiplication $\boxed{\phantom{0000000000000000000000}}$

Distributive Property $\boxed{\phantom{0000000000000000000000}}$

**17.** If Jeannie eats 1,840 calories a day, how many calories will she have eaten after 182 days?

_____ calories

**18.** There are 8 teachers going to the science museum. If each teacher pays $15 to get inside, how much did the teachers pay?

$ _____

**19.** Select other ways to write 50,897. Mark all that apply.

(A) $(5 \times 10{,}000) + (8 \times 100) + (9 \times 10) + (7 \times 1)$

(B) $50{,}000 + 800 + 90 + 7$

(C) $5{,}000 + 800 + 90 + 7$

(D) fifty thousand, eight hundred ninety-seven

**20.** For numbers 20a–20b, select True or False.

**20a.** $55 - (12 + 2)$, value: 41    ○ True    ○ False

**20b.** $25 + (14 - 4) \div 5$, value: 27    ○ True    ○ False

**21.** Tara bought 2 bottles of juice a day for 15 days. On the 16th day, Tara bought 7 bottles of juice.

Write an expression that matches the words.

**22.** Select other ways to express $10^2$. Mark all that apply.

(A) 20

(B) 100

(C) $10 + 2$

(D) $10 \times 2$

(E) $10 + 10$

(F) $10 \times 10$